Environmental Footprints and Eco-design of Products and Processes

Series Editor

Indexed by Scopus

This series aims to broadly cover all the aspects related to environmental assessment of products, development of environmental and ecological indicators and eco-design of various products and processes. Below are the areas fall under the aims and scope of this series, but not limited to: Environmental Life Cycle Assessment; Social Life Cycle Assessment; Organizational and Product Carbon Footprints; Ecological, Energy and Water Footprints; Life cycle costing; Environmental and sustainable indicators; Environmental impact assessment methods and tools; Eco-design (sustainable design) aspects and tools; Biodegradation studies; Recycling; Solid waste management; Environmental and social audits; Green Purchasing and tools; Product environmental footprints; Environmental management standards and regulations; Eco-labels; Green Claims and green washing; Assessment of sustainability aspects.

More information about this series at http://www.springer.com/series/13340

Ivan Coste-Manière · Miguel Ángel Gardetti
Editors

Sustainable Luxury and Jewelry

Springer

Editors
Ivan Coste-Manière
Skema Business School
Sophia Antipolis, France

Miguel Ángel Gardetti
Center for Studies on Sustainable Luxury
Buenos Aires, Argentina

ISSN 2345-7651 ISSN 2345-766X (electronic)
Environmental Footprints and Eco-design of Products and Processes
ISBN 978-981-16-2453-7 ISBN 978-981-16-2454-4 (eBook)
https://doi.org/10.1007/978-981-16-2454-4

This Springer imprint is published by the registered company Springer Nature Singapore Pte Ltd.
The registered company address is: 152 Beach Road, #21-01/04 Gateway East, Singapore 189721, Singapore

Preface

Centuries ago, and still today, the driving force to buy jewelry was not merely reduced to its ornamental function. Jewelry was, above all, an expression of wealth, legitimizing the social class and power of its privileged owner or a tribal symbolic meaning from an anthropologic definition. Nowadays, besides being an investment in terms of its value that increases over time, the goal when buying jewelry is pure aesthetic pleasure; even though, undoubtedly, the sign of unequivocal distinction and superiority of its owner still remains.

When asked about what brings jewelry closer to the rest of the luxury products and what brings it apart, Miriam of Hungary, Princess of Turnovo, jewelry designer, and appraiser, said: "*.... I think that jewelry is the luxury product for excellence and, in this regard, I believe I would put the question just the other way around: What brings the rest of the luxury products closer to jewelry? When a product –which is more or less necessary– is conceived, designed and executed from the luxury concept, it may come closer to the luxury essence, which is what jewelry represents... A bag, a garment, a cooking recipe, a car... they are all concepts that are based on needs and that can be materialized as luxury objects as well. However, I think that jewelry represents luxury for luxury, it is unnecessary. It is a luxury in itself. This nuance is what I think brings it closer and apart at the same time...*" (in Girón 2010, p. 69).

In the age of sustainability, we increasingly see how designers and consumers begin to think beyond a product look&feel and its function, and are especially concerned about what has happened during its manufacturing process and what will happen once its useful life comes to an end. Today, consumers value that every industrial product and process are sustainable, beneficial for the people, the economy, and the planet, and this is also the case for jewelry.

This volume[1] gives a comprehensive outlook about this subject and begins with the work titled *Encoding Values and Practices in Ethical Jewellery Purchasing: A Case History of Italian Ethical Luxury Consumption* by Linda Armano and Annamma Joy. In this chapter, they illuminate an ethical jeweler's practices and values, contextualized in the socio-cultural traditions unique to Italy through ethnographic interviews with the owner of Gioielleria (Jewelry) Belloni in Milan—the most popular ethical jewelry store—, and a representative sample of his clientele.

The following chapter, *Disrupting the Chain: The Luxury of Craftsmanship* written by Dr. Shaun Borstrock explores three London jewelers who approach design and the making process in two distinctively different ways. This chapter analyzes their shared understanding of materials and how they are used, the impact they have on the environment, and a deliberate attempt to reduce waste, adopt sustainable practices, and enhance the customer experience through hands on engagement with their products.

Then, Sudeep Chhabra and Ivan Coste-Manière develop the chapter titled *Indian Luxury Jewellery—Going #VocalForLocal*. Their research focused on the launch and commercial success of "Sabyasachi Jewellery" as an independent voice that brought the craftsman center stage. Their aim was to discover if it is possible to create an ideal balance between commercial success and sustainability of the craftsperson. Their primary method was desk research as well as interviews with customers and independent craftspersons across the jewelry industry in India. Meeting Sabayasashi himself, the king of the Haute Couture in India, has been a smashing opportunity.

Subsequently, the purpose of Annette Condello's *Viable Pearls and Seashells: Marine Culture and Sustainable Luxury in Broome, Western Australia* is to explore the origin of the hidden sea pearls and seashells as underwater luxury goods from a global-historical and marine-cultural perspective within the sustainability debate. In tracing what constitutes "sustainable luxury", considering both natural and imitation items, the chapter refers to writings concerning jewelry ethnography and more broadly the impact of the connections between the pearl, fashion, and tourism industries in Western Australia.

Moving on to the next chapter, *The Pearls and Their Soul*, the authors—Justine Ducrocq, Marion Fossati, William de Marsangy, Ivan Coste-Manière—present, on the one hand, the pearls quality approach and typology and, on the other hand, describe the bioecological characteristics of the different species of pearl oysters thanks to the advances and sustainable innovations that have made their cultivation possible. Besides, the different myths represented by pearls since their discoveries will be highlighted with the help of one of the authors who has been awarded by the Gemologic Institute of America.

[1] This volume is an effort of the **Global Center on Sustainable Luxury**, created by the agreement between **SKEMA Business School** (France) and **The Center for Studies on Sustainable Luxury** (Argentina) on July 24, 2018.

The following chapter titled, *Traceability, Sustainability, and Circularity as Mechanism in the Luxury Jewelry Industry Creating Emotional Added Value*, by Danielle Keller, investigates the luxury jewelry industry from a broadly critical perspective, one of sustainability. By researching this topic from different angles, one indeed gets a wide view of the industry. Based on this research and as one of its outcomes, a new and innovative framework for this industry's operations as a sustainable player is presented.

Then Florent Vincent, Ivan Coste-Manière, and Marc Basseporte develop the chapter titled *Tanzanites: The Maasaï Sustainable Dilemma for the Rarest Gemstones*. The objective of this chapter is to demonstrate that Tanzania is a resourceful country in terms of culture and history. This African nation should be known as a country that is being able to manage efficiently and respectfully the process of mining and trading tanzanite to enforce its local development. Starting from the global context of tanzanite mining, its history, and background, this study keeps on analyzing the structure of stones distribution. It furthermore showcases the different improvements that have been made to increase the safety of miners as much as the transparency of tanzanite market and provide work opportunities for Tanzanian inhabitants. Finally, by providing a global understanding of the current situation of tanzanite business, this research presents foresights to improve the future of tanzanite trade and establish efficient strategies to secure the market.

Subsequently, the purpose of Veronica Manlow's *Jewelry Design in the Luxury Sector: Artistry, Craft, Technology and Sustainability* shows how the creative process, work environment, conditions under which manufacturing takes place, the creations of jewelry designers, and their own experiences within the luxury trade in large global firms differ from that of those who are independent entrepreneurs or who work for smaller brands.

Moving on to the next chapter, *Luxury and Sustainability: An Experimental Investigation Concerning the Diamond Industry* the authors—Matteo de Angelis, Cesare Amatulli, and Silvia Petralito—present a research focused on a new sustainable alternative: the lab-created diamond. The technology behind this man-made stone allows companies to obtain diamonds with the same physical, chemical, and aesthetical properties of mined diamonds. In particular, the authors researched on how certain attributes specific to diamonds (namely perceived scarcity and authenticity) are perceived by consumers.

Then, Ismail W. R. Taifa develops the chapter titled *Sustainable Industrialisation for Luxury Products: Manufacturers and Retailers Must Commit to Tackling Modern Slavery in Africa*. This study focuses on Africa regarding manufacturing and retailing of luxury products. Africa is focused because the 2018 Global Slavery Index ranked Africa number one concerning modern slavery issues (MSIs); several African countries produce precious (valuable) metals; and, many African societies cannot notice much about how critical the MSIs are. The production processes and other sustainability issues were thus explored. The findings suggest the need for Africa to strengthen consolidative interventions to fight the diverse environment that results in MSIs.

Finally, Anil Kumar Bagathi, Carmelo Balagtas, Sai Vijay Kumar Boppana, Ivan Coste-Manière, Florent Vincent, François Le Troquer, and Gérard Boyer in the chapter titled *Lab-Grown Diamond–The Shape of Tomorrow's Jewelry*, intent to add value to existing researches made, link those who are interested to those who already identify themselves as enthusiasts, and later hopefully create a bond to continue this story-telling that started all the way from the seventeenth century. Developments are proactively done to reconsider what has been known for a long time and accommodate what is new. Even though it's only going to take time for the world to be convinced to believe in the integrity of an alternative: a fine commodity, this research may spark curiosity to discover the new trend that is certainly taking place in front of our eyes called the Lab-Grown Diamond.

Sophia Antipolis, France Ivan Coste-Manière
Buenos Aires, Argentina Miguel Ángel Gardetti

Reference

Girón M E (2010) Inside Luxury. London, Lid Publishing Ltd.

Contents

About the Editors

Dr. Ivan Coste-Manière is the Program Director of Luxury & Fashion Management at SKEMA Business School. His specialties include marketing, communication and international luxury goods management. He has managed the R&D of several multinational groups, and founded several companies in the areas of cosmetics and perfumes, luxury goods, PR and events, and pharmaceuticals. He has been a section member of the French Economic and Social Council, and Deputy Mayor of the city of Grasse for ten years. He has been President, with Prince Albert II of Monaco, of the Track and Field Association: Celebrities for Sports and Charities, while co-chairing the Center for Sustainable Luxury, and works as a consultant for several well-known luxury brands. He holds a Ph.D. in Chemistry, is an Engineer of the Ecole Centrale Marseille, and has been awarded the Gold Medal (Ministry of Youth and Sports) and the Officier des Palmes Académiques Medal by the Ministry of Research and Education. He is member of the Economic, Social and Environnemental Council Region Sud, of the board of the Comité Français Pierre de Coubertin, Vice Président of the Association Francophone des Académies Olympiques and a distinguished speaker in numerous countries and prestigious institutions and conferences.

Miguel Ángel Gardetti (Ph.D.) founded the **Centre for Study of Sustainable Luxury**, first initiative of its kind in the world with an academic/research profile. He is also the founder and director of the "Award for Sustainable Luxury in Latin America". For his contributions in this field, he was granted the "**Sustainable Leadership Award** (academic category)," in February, 2015 in Mumbai (India). He is an active member of the **Global Compact** in Argentina –which is a **United Nations** initiative-, and was a member of its governance body –the Board of The Global Compact, Argentine Chapter - for two terms. He was also part of the task force that developed the "**Management Responsible Education Principles**" of the United Nations Global Compact. This task force was made up of over 55 renowned academics worldwide pertaining to top Business Schools. He has several publications in the sustainable luxury arena.

Encoding Values and Practices in Ethical Jewellery Purchasing: A Case History of Italian Ethical Luxury Consumption

Linda Armano and Annamma Joy

Abstract The concept of ethical jewellery today includes certified Fair mined gold as well as conflict-free and ethically mined diamonds; consumption of such products is soaring in tandem with evolving consumer interest in ethical labour practices and sustainability. While previous studies investigate ethical jewellery consumption, the literature is silent on such consumption by Italian consumers. We aim to close this gap through a case history of Italy's first and currently most popular ethical jewellery store, Gioielleria (*Jewellery*) Belloni in Milan. Through ethnographic interviews with the owner and a representative sample of his clientele, we illuminate an ethical jeweller's practices and values, contextualized in the socio-cultural traditions unique to Italy. We further examine how Italian ethical consumers interpret Canadian ethical diamonds and certified Fair mined gold during their purchasing experiences, vis-à-vis their perceptions of the Made in Italy brand. We employ the ethnopragmatic theory [23] and the concept of territorialization [19] in our conceptual framework.

Keywords Ethical jewellery consumption · Italian ethical jewellery consumers · Conflict-free diamonds · Fair mined gold · Made in Italy brand · Ethnopragmatic theory · Territorialization

---

The authors gratefully acknowledge receipt of a Social Sciences and Humanities Research Council grant on luxury SSHRC 435-2013-1211. This project has also received funding from the European Union's Horizon 2020 research and innovation programme under the Marie Sklodowska-Curie Grant Agreement no. 837190. The authors acknowledge the receipt of a Marie Curie Fellowship to Dr. Linda Armano.

L. Armano · A. Joy (✉)
University of British Columbia Okanagan, 3333 University Way, Kelowna, BC V1V 1V7, Canada
e-mail: annamma.joy@ubc.ca

L. Armano
e-mail: larmano@ubc.ca

L. Armano
Università Ca' Foscari Venezia, 30123 Venezia, VE, Italy

I. Coste-Maniere and M. Á. Gardetti (eds.), *Sustainable Luxury and Jewelry*,
Environmental Footprints and Eco-design of Products and Processes,
https://doi.org/10.1007/978-981-16-2454-4_1

1 Introduction

The concept of ethical jewellery today includes certified Fair mined gold, along with conflict-free and ethically mined diamonds. Consumption of ethical jewellery has soared in recent years, in tandem with rising consumer interest in fair labour practices and environmentally sound manufacturing. The issue of ethics in jewellery production has gained equal weight to sustainability; whether relating to global mining activities [42], supply chain management [48], or labour practices, ethics has risen to the forefront of consumer consciousness. Non-profit organizations, activist groups, and international governmental institutions have drawn consumer attention to ethics in jewellery manufacturing via campaigns addressing corporate social irresponsibility [34, 35, 46], including lack of transparency, human rights abuses such as child labour, environmental degradation from mining, and the role of conflict diamonds in funding terrorism. These criticisms present fundamental challenges to the organizations of the global jewellery supply chain, as well as corporate governance [47]. Many scholars [39] note that most of the major mining companies and luxury groups have, over the past 20 years, embraced ethical and sustainability positions; Tiffany and De Beers have long cited social and environmental responsibility as intrinsic to their business models. Accordingly, consumer criteria for fine jewellery purchase decisions are now likely to include ethical business practices [47], although research in this area remains scant [46].

Most approaches to the issues of luxury consumption start from social theory and are often based on an ontological perspective of dualism between the individual consumer and the overall social structure. This dualism has been challenged by alternative sets of understandings of human action in context [49, 54], which traverse individuals and structures, and assume a societal framework comprising social practices produced and reproduced across time and space [28]. These different approaches can be inserted in the common category described as practice theories [46]. While practice theories are employed in a diverse range of topics related to consumption and marketing (Echeverri and [24], their primary application relates to research on consumption [2, 31, 67] and studies on behaviours modified to reflect a commitment to sustainability [25, 50, 51, 57].

While economic models assume that human actions are dependent on individual objectives, and other social theories tend to see actions as a form of compromise between social norms and the need for some degree of consensus, practice theories seek to highlight action through symbolic processes of interconnected meanings [49].

The proliferation of approaches attempting to overtake the implicit dualism in the social science literature shares some commonalities whereby practice explanations are related to human activities, knowledge, meaning, language, social institutions, and historical transformations involved in the field of practice [55]. In many studies consumers are visualized as practitioners rather than authors of their own actions; thus, they can execute actions according to a shared social background of knowledge and understanding of normality, which in turn are interpreted subjectively [25]. This conception implies that different forms of consumption are embedded in the social underlayer and linked to individuals' judgments of what comprises normal actions [50, 57].

Despite the significant corpus of works on consumption, no clear, unified practice theory approach exists in extant literature regarding ethical and sustainability issues in consumption studies, despite the urgent timeliness of understanding these issues [26]; indeed, the very definition of such issues is unclear. For example, some investigations suggest that consumers' meanings do not consider the potentially negative impact of luxury consumption, since luxury products do not appear frequently in commodity consumption practices. Moreover, consumers may perceive such products as unaffordable, thus rendering the ethical issues related to such goods moot [17].

As suggested by Moraes et al. [46], ethical issues in luxury consumption need new processes to understand how people incorporate materials, knowledge, and meanings in their ways of consuming. Moraes et al. suggest applying practice theories, not only to study routine purchases but also to examine non-habitual practices such as ethical luxury consumption. Notably, they further propose linking norms and the consumption environment to consumers' behaviours. Via this approach, scholars aim to explore how these practices are learned and shared among consumers in their daily lives [46].

While previous studies investigate ethical jewellery consumption, the literature is silent on such consumption by Italian consumers. In the present study, following Moraes et al., we aim to close this gap, through the lens of a case history of Italy's first and primary ethical jewellery store, Gioielleria (Jewellery) Belloni in Milan. Through ethnographic interviews with owner Francesco Belloni and a representative sample of his clientele, we illuminate the specifics of an ethical jeweller's practices and values, contextualized in the socio-cultural traditions unique to Italy.

Rather than rely on a theory of social practices applied to studies of luxury consumption, we furnish a completely new conceptual framework, applying the ethnopragmatic theory [23] and referred to hereafter as ET, which has heretofore never been applied in an analysis of luxury consumption. Through our methodology, we examine how Italian ethical consumers interpret Canadian ethical diamonds and certified Fair mined gold during their ethical purchasing experiences, including through comparisons with their perceptions of the Made in Italy brand. To deeply understand the overlapping consumption spaces of Canadian jewellery products on the one hand and their Made in Italy counterparts on the other—an overlap unique to Italian consumers' experiences–we further introduce, for the first time in a study of consumption experiences, the concept of territorialization [19].

2 Methodology

In terms of a theory of social practices, we not only include norms and consumption environment to promote behaviours [46] but also unconscious actions [8] and cultural values that locate habits in a particular cultural environment. While Bourdieu

(1997) emphasizes the importance of embodied habits for human actions, we follow Akrivou and Di San Giorgio [1] in considering human habits not only as reproductive actions of an existing status quo, but also as a dialogical conception, which allows for new gradual co-creations and co-evolutions of novel actions emerging between acting agents (e.g., jeweller and clients). This dialogical perspective is particularly applicable to ET [23] and the concept of territorialization [19] in our luxury consumption analysis. Duranti [23]'s methodology, which represents the theoretical concerns driving our research approach, gives pragmatic force to language, with insights taken from post-structuralist social theorists such as Anthony Giddens [27, 28], Pierre Bourdieu [8–10], and Charles Tylor [61], as well as phenomenologists like Edmund Husserl and Emmanuel Levinas (Duranti 2009a). Duranti, referring to Bourdieu's concept of "habitus", also specifies that in the analysis of the communicative performance between subjects, not only the said words must be considered but also the unconscious behavioural aspects (of which people are often unaware) must be observed and interpreted, which are in any case effective in transmitting the agentive entities related to communication and understanding of the aesthetic, moral, and practical contents transmitted through language.

Thus, Duranti perceives the world as a context full of meanings [23] in which subjects interpret their surroundings according to hermeneutic practices shaped through countless previous normative values. Specifically, he asserts that in most cases people do not have conscious control over their interpretations of the world, even if their interpretations, as well as values and behaviours, are limited.

In the present study, Duranti's ET is adapted to encode cultural values, as unconscious and conscious thoughts emerged during our interviews with the jeweller and his clients. By adapting ET, we can compare our participants' comments to highlight their discrete intentions to create common interpretations of ethical jewellery, Fair mined certified ethical gold (FCEG), and Canadian ethical diamonds (CED), and their perceptions of unethical versus ethical mining activities.

In the present research, we cannot ignore the important aspect of intersubjectivity, which we use to understand both the relationship between subjects in the retail store, in which the people interviewed evolve to experience direct sensory access to each other, indirect access with imagined subjectivity (such as miners employed in the mines in other parts of the world and local communities near the mines), and a connection, acquired through personal experiences and media exposure, between consumers and an imagined environment related to mining activities.

Inevitably, perceptions of what comprises ethical jewellery given by our participants embed global advertisement narratives that frame CED and FCEG, whose influence we must consider. Nevertheless, we do not interpret transnational knowledge linked to minerals as a way to impede people's ability to be reflective in the context of commodities regimes. Quite the opposite—considering the wide advertisement discourse, we must both highlight and contrast the cultural flexibility of interpretation that allows our participants to reshape the global messages of CEDs and FCEG as applied to Italian fine jewellery. This process of malleability is concretized through the comparisons of participants between their cultural comprehension of the idea of Made in Italy products (which are perceived as a familiar concept) and ethical jewellery.

3 Ethnographic Research

Our research aims to provide a deeper understanding of how Francesco Belloni engages ethical clients who are exhorted to contribute, thanks to their virtuous purchases, to save the planet and defend particular categories of people (such as workers and children) employed in mining activities in other parts of the world. Our work is informed by the salient elements of an ethnographic account [37, 38] via in-depth interviews with the jeweller and 15 ethical clients, both male and female from diverse parts of Italy, and systematic data collection gathered through the integration of oral and written documentation on mining activities, as well as on Fairtrade certification schemes, supply chain management linked to ethical diamonds and gold, and various case histories of ethical purchases in fine jewellery contexts (Le Billon 2008; [29, 46]. A number of important documents were collected at The Prince of Wales Northern Heritage Centre archive in Yellowknife in Canada's Northwest Territories in order to deeply examine the economic and historical process of ethical certification and global narrative discourses linked to Canadian diamonds. We used a phenomenological approach [63] combined with ET in our interviewing process with a questionnaire at the end of the ethnographic research, which allowed us to summarize the most important emergent topics. All interviews lasted from 60 to 90 minutes, were recorded with the permission of interviewees, and focused on individual experiences. Participants were chosen on the basis of their consumption and frequency of visiting Gioielleria Belloni and their knowledge of CEDs and FCEG. Participant ages ranged from 27 to 40; most were affluent.

Thanks to the methodology adopted during the interviews, subjectivity was a key component in eliciting the deeply embedded meanings of ethical purchases, both in consumer experiences and in the linguistic ways deployed to explain them, to draw out participants' otherwise untapped responses. Among all interviews, the commonly asked questions included: What does the concept of ethics linked to diamonds and gold mean to you? When and why did you choose to buy ethical jewellery instead of other kinds of jewellery? What does being an "ethical consumer" mean to you? How do you imagine mining activities in the Northwest Territory of Canada where ethical diamonds come from? How do you imagine mining activities in artisanal and small mines linked to Fair mined certification? And lastly, how do you perceive mining activity? All interviews were transcribed and analysed to identify emergent narratives and cultural values (Table 1).

During our ethnographic research, we synthesized existing narratives about CEDs and FCEG (from literature as well as mining companies' and luxury brands' reports and websites) in nodes emerging from the findings. In accordance with Sippl [58], we found that commodity regimes targeting minerals include a range of transnational initiatives varying according to different categories such as minerals (gold,

Table 1 Participants divided into age classes (from 25 to 40 years old; and from 40 to 60 years old)

Participants from 25 to 40 years old	Participants from 40 to 60 years old
Marco (37, Office worker, Genoa); Filippo (33, Banker, Varese); Sara (37, Doctor, Brescia); Francesca (34, Art Director, Rome); Giacomo (32, Geologist, Padua); Laura (37, Office worker, Pistoia); Giada (37, Office worker, Lecce); Pietro (37, Engineer, Milan);	Francesco (59, Jeweler, Milan); Francesca (42, Office worker, Milan); Manuela (42, Manager, Torre del Greco); Mauro (51, Doctor, Milan); Carlo (42, Office worker, Piacenza); Letizia (40, Office worker, Bollate-Milan); Umberto (50, Architect, Milan); Luigi (45, Doctor, Rome);

diamonds, tin, and tungsten); issues of interest (conflict, environment, poverty, and worker rights); producer scale (large versus small or artisanal scale); program type (third-party certification, building projects, and second-party labelling); and program duration (one-off grant-funded pilots versus ongoing, self-sustaining initiatives) [58]; Van [64]. To better understand how the global narratives on CEDs and FCEG interacted with Italian ethical consumers' discourses, we focused on "minerals" (gold and diamonds); "issue of interest" (conflict, worker rights, environment); "producer scale targeted" (large scale in Canadian context; small or artisanal scale in South countries mines); "program type" (traceability, referring to the physical tracking of minerals at all points on the supply chain, from the mine of origin to the customer, and certifications), and "imagined geographic dimension" as explicated in the media (Africa versus Canada; polluting small and artisanal mines in undefined places in the world versus sustainable small and artisanal mines, thanks to the Fair mined program).

A key difference between "minerals" and "issues of interests" is their respective differing global narratives: diamonds versus conflicts and human rights, and gold versus environmental pollution and labelling programs. "Blood diamonds" are portrayed as funding violence and wars (Le Billon 2006; [16], and are linked to criminal projects, motivated by lust for resources and viewed as an easily accessed loot source ([15, 40], with warlords [18] viewed as exploiters of men, women, and children subjected to widespread abuses (Le Billon 2008).

The first guarantee encourages consumers to have an ethical perception of diamonds, and is linked to our category of "producer scale targeted" which, in the case of Canadian diamonds, is associated with large-scale operations. As some participants noted, industrial exploitation is associated with stronger economic growth: a poor human-rights record directly equals lower prices. Specifically, the exploitation of primary or kimberlite requires industrialization [kimberlite is an igneous rock, which may feature diamonds]. In contrast, secondary or alluvial deposits, often found in poorer countries, are open to artisanal mining with or without the official consent of authorities (Le Billon 2008). Privileging industrial over artisanal mining has significant political consequences. Compared to artisanal exploitation, industrialization concentrates power in the hands of the state that, in the case of Canada, globally certifies rhetoric on ethical employment opportunities and reflects the preferences of foreign extractive companies obligated to respect Canadian laws. Thus, diamonds can be perceived as emblemizing Canada itself [3] in which social issues are the platform not only to sell products but also to sell the Canadian Government brand. In this context, CEDs are also related to the category of "program type" through certification of origin. Indeed, the government of the Northwest Territories of Canada is the first in the world to issue government-certified certificates of authenticity to diamonds that are mined, cut, and polished in the Northwest Territories. The Canadian brand is globally presented as signifying the purity of the diamonds, both geologically and politically. This discourse proposes a Canadian diamond mine's characterization as, in environmental terms, incorporating "clean mining" and, primarily, embodying social responsibility. However, the actual process of Canadian diamond certification is somewhat nebulous, in that not all diamonds from Canada carry certification that is from Canada. In order to state that such diamonds are Canadian in origin, their certification should be traceable all the way from the rough diamond to the wholesaler's office and finally through to the retailer who sells the diamond. In order to certifiably be from Canada, the diamond has to have a logo for the Canadian Diamond Code of Conduct and the seller must be listed as a member of the Canadian Diamond Code of Conduct. (https://purediamond.ca/info/how-to-make-sure-your-diamond-is-canadian/).

In its response to the "blood diamonds" discourse surrounding African diamonds, Canada has been the leading country in proposing a certification scheme, launched in 2003, that surpassed the Kimberley Process (KP) plan, a certification scheme itself, initiated in 2002, to prevent blood diamonds from entering the market for rough diamonds [7]. The KP can be fairly represented as the project preceding the Canadian diamond certification in the fight against conflict diamonds. For decades, the illegal process of mining and marketing rough diamonds was accepted by mining companies. In 2000, however, De Beers announced that they were no longer willing to trade diamonds from conflict zones, particularly from African mining contexts. Governments and mining industries began to work together to combat the illegal diamond trade and, by founding the World Diamond Council, they signed the KP plan. In contrast to Canada's ethical diamond certification of origin, the KP does not allow consumers to know the exact origin of a given diamond, but simply guarantees, thanks to the signature of governments, mining companies, and other for-profit and non-profit organizations, as well as of private citizens, that a specific diamond entering the market has not been derived from illegal trafficking managed by criminal organizations [32].

Thanks to a global marketing strategy of the mining companies operating in Canada, the opposition is automatically created between state-regulated Canadian mines and African artisanal mines: two opposite peaks in the geopolitical landscape of diamond mining and trade (Le Billon 2008). Schlosser [56] argues that CEDs, presented as the pure alternative to African "blood diamonds", often constitute a crude attempt to establish linear links between different places that invoke poverty and natural resources (D'Angelo 2019) as well as images of Africa as synonymous with violence and primitivism (Le Billon 2006)—in direct opposition to symbolic imagery of northern Canada as a pristine and implicitly moral rural space white with snow.

The global narrative regarding artisanal and small-scale mining centres is on health hazards, particularly because mercury amalgamation in gold mining can precipitate human and environmental damage. As numerous studies show, in addition to air and water pollution due to cyanide and mercury (used for the sorting of ore from waste rock), the environmental degradation caused by the mine also compromises the livelihoods of those living near mining areas [20]. Moreover, environmental pollution raises issues of land ownership conflicts involving miners, collective or private concession holders, and indigenous communities [20].

Discourses on artisanal and small-scale mines create images implying that gold mining directly leads to or indirectly financed armed conflicts over the mineral. This representation is supported by many studies on African artisanal mines [20, 45] in which mined diamonds finance rebel groups (d'Angelo 2019; [66]. As of this writing, distinctions between "artisanal and small-scale mining" are undefined and often conflated in the literature. Many scholars [18, 44, 45] affirm that artisanal and small-scale mining (ASM) in combination are largely an informal sector with limited available information on production, revenues, operations, and even the location of activities. Empirical evidence supporting formalization of this sector as a means to improve working conditions (e.g., by the Minamata Convention, a global treaty designed to protect human health and the environment from the adverse effects of mercury use in mining activities) is largely unavailable to researchers. We do know that artisanal mining activity is typically informal, lacks sophisticated machinery, and attracts destitute workers. In contrast, small-scale mining activity is more organized, although miners do not necessarily use sophisticated machinery even if they have higher revenue turnover and legal licenses.

Over the past 30 years, both small-scale and artisanal mines have increased globally [20], yet as of this writing, no existing agreement fully defines and regulates either type of mines, a situation complicated by the fact that what regulation there is varies from state to state, according to the internal politics and economy of each nation.

In our category of "producer scale targeted" linked to artisanal and small-scale mines, the general representation of artisanal mines depicts scenarios in which small groups of unorganized workers, often within families, are involved in mining, while the representation of small-scale mines supports narratives of mining management through small cooperatives. Drawing on existing literature, however, both types of mining organizations can be regulated legally or illegally [33, 36].

In the 1990s, governments in Columbia, Peru, and South Africa began promoting an image of environmental sustainability. Given its role in tackling the degradation caused by mining activities and providing supply-chain tracking guaranteeing fair labour and environmental policies, such sustainability represents an alternative to environmentally damaging mining activities. The Fair mined program aims to formalize artisanal and small-scale mines into a profitable activity using efficient technologies, end child and labour exploitation, and reduce poverty in participating countries. Beginning in 2009, artisanal and small-scale mines worked with the Fairtrade and Fair mined gold [14, 62]. Starting in 2013, the world standard for defining mined gold, silver, and platinum evolved from Standard Zero (the world's pioneering standard for mined gold, silver, and platinum) to Standard 2.0 (created in 2014 and involving stakeholders throughout the supply chain). A third-party certification program checks compliance with the standard for artisanal and small-scale mines every year. Today, 12 artisanal and small-scale mines worldwide are certified as Fair mined (two in Peru, two in Bolivia, seven in Colombia, and one in Mongolia). Certified artisanal and small-scale mines receive at least 95 percent of the gold price set by the London Bullion Market Association, which is higher than the usual market price (European Commission 2010). In addition, miners receive a Fair mined premium of USD $4,000 per kilogram [41]. Taken as a whole, the imagined geography of symbolic environmentally friendly mines, and of miners working in safe conditions with Fair wages, is further entrenched in consumers' minds.

3.1 Gioielleria Belloni: Contextualization of Ethnographic Research on Italian Fine Jewellery Purchases

Gioielleria Belloni has a venerable history dating to 1926, when the grandfather of the current owner, Francesco Belloni, opened the store in Milan. With Francesco at the helm, Gioielleria Belloni first engaged in charitable activities in 2002, donating part of the proceeds from their jewellery sales on Valentine's Day and Christmas to non-profits active against cancer and supporting minors in poverty; moreover, 10 percent of the customer jewellery purchases were donated to various charitable organizations. These donations were always communicated locally, through regional television networks, or in radio broadcasts, often in advance.

In 2003, Francesco attempted to donate to the Londoner Survival International (a human rights organization formed in 1969 whose campaigns support human rights in developing countries); the organization refused the donation based on Belloni's connection to diamond sales. The group had already been boycotting the Diamond Trading Company and De Beers in Botswana for several years (https://www.survivalinternational.org/); the latter company had expropriated the Bushmen [hunter-gatherers indigenous to Southern Africa] from their land, especially in the Kalahari Desert, where diamond deposits had been found.

At the time, Francesco was sourcing African diamonds only. Survival International, in response to his plea, provided an alternative: diamonds from the Ekati and Diavik mines in the Northwest Territories of Canada. In 2005, Gioielleria Belloni began selling CEDs in Italy; the company's website terms Fair mined gold "The gold to be proud of". A grateful clientele immediately perceived Canadian gemstones as free from illicit trafficking, unlike African diamonds. Francesco specifically noted that the Canadian origin of his diamonds was a key selling point to his customers, to the point that he identifies his clientele as ethical customers, as opposed to other customers who choose not to purchase ethically produced jewellery. Francesco created the Ethical Diamonds brand, whose logo is a stylized diamond featuring a red maple leaf, which he has promoted widely online and through advertisements on Radio Popolare di Milano, a station with a politically liberal viewpoint whose listeners tend to embrace environmental sustainability and ethical labour practices.

The Ethical Diamonds brand focuses on supply chain certification. As Francesco noted during our interview with him, the Gioielleria Belloni website quickly became a major e-commerce source, both locally and worldwide for ethical Canadian diamonds. Before introducing CEDs, the jewellery shop sold around 60 diamonds a year; in recent years, that number has hovered around 300, specifically of CEDs. In 2008, Gioielleria Belloni also started selling FCEG from artisan mines in Columbia and Peru, thus creating the "Ethical Gold" brand. Francesco buys pure gold in grains and sends it to workshops outside the shop that process it, ensuring that certified Fair mined gold is separated from its non-certified counterpart.

4 Encoding Values: How Italian Ethical Consumers Perform Jewellery Purchases

Applying ET allows us to study *agency* through two different lenses: performance and encoding [21, 22]. Close attention to the performance of language suggests that even before constituting specific speech acts, the use of language affirms the speaker as well as the listener as potential agents. Thus, we apply the performance of language to our analysis of interviews to highlight how Francesco and his clients co-create a discursive universe on CEDs and FCEG, as well as on ethical mining activities and ethical consumption. In adapting Duranti's encoding theory in our analysis of interviews, we substituted an ethnographical codification of values and representations of Duranti's original linguistic application on the study of grammar. Duranti's explication about the ergative-absolutive type of language (EATL) inspired us to construct a frame able to encode ethical purchasers' values. Notably, the key assumption we derived from EATL is the dependency of the agent's identity (i.e., jeweller and clients), as well as their values and behaviours, on the object (i.e., ethical jewellery).

Premising that CEDs and FCEG (as objects driving agency in Italian fine jewellery consumption) establish the background of our analysis, we summarize in Table 2 below data derived from our interviews: (1) the main values that encourage ethical clients to choose ethical jewellery; (2) the ideological representation of mining activity; and (3) the cultural representation of ethical mining activities.

As illustrated in Table 2, Gioielleria Belloni clients have a clear perception of both the values that guide them in their ethical jewellery purchases, and of what comprises an ethical mine. Starting from a quantitative point of view, Fig. 2 shows that the number of participants in the two first columns is higher than the third column—many people interviewed gave two or three explications in the same response—while

Table 2 Data derived from our interviews with participants

Main values able to guide clients toward ethical choices	Ideological representation of mining activity	Ideological representation of ethical mining activity
Human rights: 15 clients interviewed	Hard work: 12 clients interviewed	Workers respected: 8 clients interviewed
Environmental impacts: 11 clients interviewed	Risky work: 7 clients interviewed	Safe workplace: 5 clients interviewed
Guilt sense: 9 clients interviewed	Underpaid work: 4 clients interviewed	Fair wages: 4 clients interviewed

as shown in the third column, just two clients gave two interpretations of ethical mines within a single reply; the remainder each gave only one interpretation. From a qualitative perspective, this information suggests that participants had not fully embodied the concept of ethical mining, nor, as hypothesized below (as shown in Sect. 5), of ethical jewellery.

This evaluation allows us to clarify how Gioielleria Belloni's ethical customers distinguish themselves through their purchase choices from store clients. Based on our data, we surmise that, while they were not fully aware of what constituted ethical mining activity, their purchases of CEDs and FCEG fine jewellery allowed for self-expression as moral individuals through consumption. Furthermore, we can hypothesize that ethical fine jewellery embodies a moral code and, as such, serves as a means for clients to experience not only their personal attitudes towards consumption but also their existence as a whole. Thus, as a result of their ethical choices, they have achieved their own integration within the larger world.

We asked Francesco the same set of questions offered to our other participants, as summarized in Table 3.

Table 3 Data derived from retailer

Main values in the jeweller's suggestions to his clients in the realm of ethical purchases	Ideological representation of mining activity	Ideological representation of ethical mining activity
Protect workers' rights linked to diamonds and gold extraction	Workers don't follow safety protocols in poor countries	Mining companies operating in Canada follow Canadian laws; therefore, workers' rights are respected Thanks to Fair mined certification we have the same guarantee for artisanal and small-scale mines
Environmental safeguards above all for large-scale industrialisation, even if artisanal and small-scale mines have less effect on the environment	Uncontrolled use of mercury in small-scale and artisanal mines	More sustainability thanks to recurring controls

In comparing the communication styles and responses of Francesco and his clients, the concept of *intersubjectivity*—between the former and the latter, as well as between our participants and their varied theorized communities of miners and geographical spaces—would appear to play a crucial role. Intersubjectivity is the common denominator in all responses creating high involvement among participants such that they become a community comprising people who share common values, experiences, and ideas, whose setting is a retail store (or its online presence). Furthermore, our participants' answers displayed a strong communicative empathy. Some clients connected characteristics related to ethical jewellery with their conception of Made in Italy products, as shown in Sect. 5.

4.1 Performance in the Discursive Universe on Ethical Jewellery

To capture our understanding of the Italian ethical luxury performance as experienced by its practitioners, we turn to the discursive universe on CEDs and FCEG as extrapolated from our interviews, to highlight the cultural values that inform Francesco's interactions with customers as he guides them towards ethical purchases. Specifically, following Duranti's assertion in which agency is enacted and represented through language, we have aggregated conceptual nuclei, derived from our interviews that articulated specific values, thoughts, and behaviours. During our conversations with Francesco and several customers, variations on the phrase "We want to be different thanks to ethical options" often emerged. Our participant Luigi, a repeat ethical client, took pride in his potential to influence the wider market: "If you are no longer buying [*non-ethical*] products, its manufacturer loses business and will perhaps change its ways. As a consumer, every time you buy something you can *make a difference*".

In Fig. 1, we show how "to be different" is converted to perceptions of ethical behaviour, for both Francesco and his customers.

Starting from Fig. 1, we build four categories in which we summarize our interview data, all precipitated by the phrase "We want to be different thanks to ethical options". These categories are arranged along two axes that comprise the two concepts of "guarantee" and "charity", around which the discursive universes on ethical jewellery of our participants were built while distinguishing Francesco's discourses from those of his clients. The concept connecting every category is "supply chain".

Compared to other case histories from extant literature [13, 46], Fig. 2 allows us to convey both the similarities and the discrete interpretations (as highlighted in the lower-left category) of ethical jewellery given by our participants. Focusing on clients' responses, our study aligns with other analyses of fine jewellery ethics (i.e., [12, 46] on the prioritization of ethical issues when consumers purchase jewellery. These issues encourage consumers to be adamant about buying only conflict-free jewellery. Furthermore, most participants mentioned the issue of "blood diamonds" as seen in various media (e.g., the 2006 film *Blood Diamond* starring Leonardo

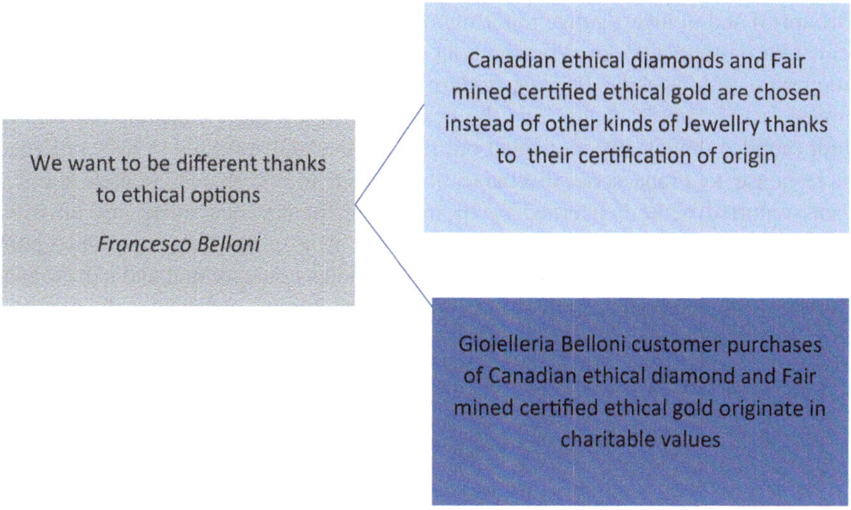

Fig. 1 Conversion of the concept « to be different » into ethical behaviours

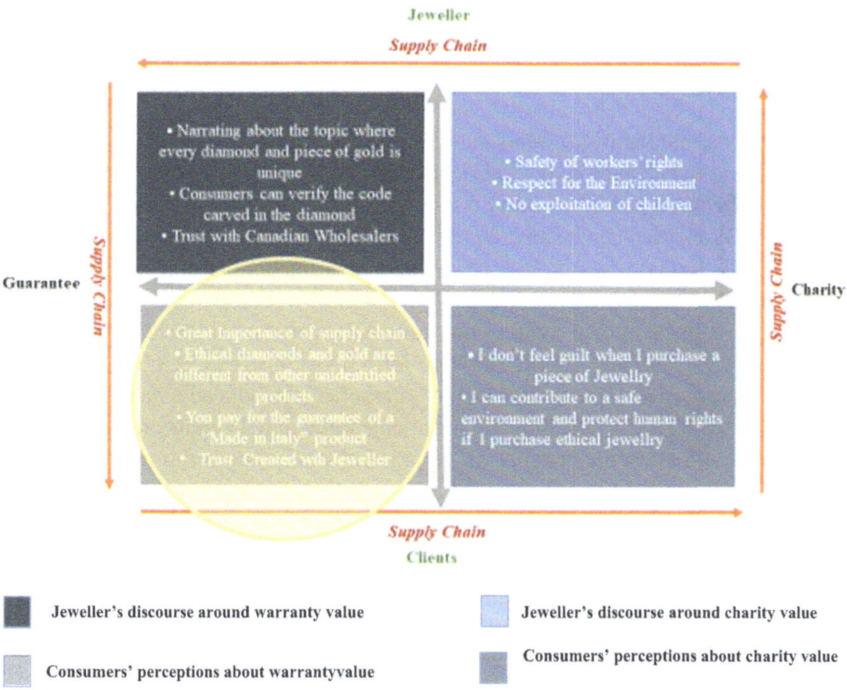

Fig. 2 Discursive universes on ethical jewellery of interviews participant

DiCaprio) and in investigative reporting on labour and human rights issues, consistent with participants' knowledge about supply chains being important in ethical purchases [46]. Specifically, our study also underscores the relationship between ethical issues, supply chains, and the possibility of jewellery consumption free of guilt (an idea that clearly motivated our participants and appeared to have emerged in response to Francesco's discourse of charity). Many of our clientele participants reported being concerned by the potential for negative social and environmental impacts from their purchase habits [43]. Consumers may experience guilt when their luxury consumption, which often denotes non-essential and extravagant spending—the embodiment of conspicuous consumption—contradicts their values. Our participant Umberto explained his choice of ethical jewellery:

> I can buy ethical jewellery. [The pieces] … cost more than other jewellery, but they… [reflect].. my point of view. I don't want to buy a diamond or gold that can damage other parts of the world. I'm lucky, and I try to share my luck around me as I can

Letizia, another participant, told us how she tries to reduce her sense of remorse by purchasing ethical jewellery:

> I think that Fair mined gives guarantees… to the workers, to local communities, and to the consumers about environmental and ethical questions. If I buy a ring or another piece of jewellery, I know that my purchase doesn't damage the environment or other people. I don't feel guilty because I want jewellery. I think that projects [such] as Fair mined can help people to live better.

As demonstrated above, clients can convert the value of "having" to a value of "being". As a result, they may experience reduced feelings of guilt that otherwise might have been triggered by luxury consumption. In this context, a charitable narrative told by Francesco can be an effective guilt-reducing strategy and can encourage brand loyalty. Extant research [59, 65] explains the impact (as shown in the above-right category in Fig. 5) of a business on implementing a perception of a cause-related campaign to support human rights or environmental causes: enhanced positive brand awareness [4, 11, 30] and a short and long-term boost in sales [59, 60, 65].

For our participants, the concept of a "guarantee" introduced a narrative around the supply chain. Both Francesco and his clients immediately connected the concept of the supply chain to the concept of "guarantee" and perceived it as indicative of the *uniqueness* linked to each piece of ethical jewellery. As Francesco reported to us, the characteristic of uniqueness is perceived as belonging only to the Canadian diamonds and to FCEG:

> Canadian diamonds are free from illicit traffic. …only Canadian diamonds can guarantee this through [the] certification of origin. For example, the Kimberley Process doesn't tell me where a diamond comes from. It is like when you go to the supermarket and you want to know where the food comes from. It is absurd that this process isn't applied to the diamond as well. Especially, for Canadian diamonds, the supply chain works well (…). Diamonds from Canada have a maple leaf carved by the laser and codes with numbers and letters. In this way, you can understand that every diamond is unique. Moreover, every wholesaler has to have a filing system. Therefore, every consumer can insert the diamond's code on a website and know who the supplier is, the characteristic of the stone, and its size before being carved. The consumers appreciate this characteristic. For Fair mined gold is different.

> I have my gold from European suppliers who buy the metal from Fair mined-certified South American mines. They sell to me pure gold in grains, and then I send it to workshops outside my shop that process it, ensuring that special care is taken to separate it from non-certified gold

The concept of "guarantee" has strong connections to perceptions of "relationship" and "trust" with two different applications. The first refers to the relationship and trust created over time between Francesco with his Canadian diamonds' wholesalers in Canada and Fair mined gold suppliers in Switzerland, Germany, and France. As Francesco reported to us: "I have a good relationship with my suppliers. I can tell them: "Can I pay you next month?", and they reply to me: "Ok". I could create this familiarity with them during the last years. (…) They trust me, as I trust them"). A sense of familiarity, and thus trust, was also experienced by Francesco's ethical clients towards the jeweller (as Luigi affirmed: "Francesco Belloni is (…) a point of reference for me"). The relationship between the jeweller and his customers is initially characterized through education, as Francesco defines the rationale for ethical jewellery. As Francesco asserted:

> My clients are usually prepared, they read the news, etc. The consumers that want ethical jewellery have to be very motivated to search for this kind of product. There are other consumers who come to the store and I explain the ethical alternative. I always explain to the consumers… about the supply chain, and this is the most important thing for them

5 Territorialization, Ethical Jewellery, and the Made in Italy Concept: How Italian Ethical Consumers Interpret Canadian Ethical Diamonds and Fair Mined Certified Gold

Figure 2 in the lower-left category shows the specific interpretations of ethical jewellery given by our clientele participants. To discuss this cultural meaning, we reclaim the concept of uniqueness, which is directly linked to the concept of differentiation between CEDs, FCEG, and jewellery produced without ethical production and manufacturing processes. The presence of a certification of origin instantly reassures consumers that a given piece of jewellery has been ethically created. Moreover, such certification engenders a higher price point, which consumers willingly pay to know the path followed by diamonds and gold from mines to purchase. Interestingly, our clientele participants were willing to pay more not only for an ethical piece of jewellery but also for a Made in Italy product. As our participant Umberto noted:

> Many people in Italy are accustomed to evaluating the quality of the products. My family and I are also very attentive to the "Made in Italy" [brand], and we always try to buy Italian products, even if often they cost more than other goods. I think that ethical diamonds and gold could be compared with Made in Italy products, where you can… know where the raw materials come from, as well as for the characteristic of costs

We can thus hypothesize that the parallelism between ethical jewellery and Made in Italy products could be an intrinsically Italian approach to conceptualizing CED and FCEG through a well-known cultural and economic concept. As emerged during other interviews, such parallelism between the categories of products also relates to consumer cultural perceptions of highly educated workers, with significant accumulation of experience. Umberto told us:

> When I think about ethical mines, I think that the workplace is safe for workers and it is a warranty for the consumers. (In the ethical mines) companies allow the worker to be safe and well paid… I think that miners (in Canada) are professional workers, chosen by the company for their skills. The company analyses their CVs, and maybe it can also test… [the workers] to…[gauge] their expertise. Then they can work in the mine… I also think that these workers can earn a good salary, better than other jobs, thanks to the Fair mined label

While the parallelism between CEDs, FCEG, and the cultural idea of Made in Italy products has never previously been studied, the concept of territorialization [19] can help us understand this parallelism. Territorialization (*appaesamento* is the original definition coined by Ernesto De Martino) denotes the opposition of *spaesamento* (which derives from the Italian word *spaesato*) that translates to "bewildered"; thus, territorialization is the exact opposite of bewildered. In our analysis, the concept of territorialization is useful in studying the stratified way in which new and old meanings are blended by our Italian clientele interviewees. The De Martino's phrase "non-problematic setting of the obvious" [19] explains how the familiarity of acts and objects that constitute our world, as well as our knowledge expressed through linguistic and non-linguistic means, allow us to gradually expand our known territory, embracing new meanings and practices. In this way, territorialization is a form of the "domestication of the world" [19], in which an individual's understanding of new acquisitions (whether of the mind (e.g., cultural or linguistic) or the physical world (e.g., the import of new products) is filtered through the tradition in which a person belongs. In this way, through the concept of territorialization (ibid.), the individual pragmatically acquires cognitive security and an emotional trust that allows for being in the world [53]. De Martino's conceptual approach to territorialization, from the German philosopher Martin Heidegger, implies the idea of building and living not only of places but also of forms of knowledge that in turn is the practical condition underlying the cultural creation of subjectivity [5]. To further explain De Martino's theorization of territorialization, it is useful to briefly retrace the theoretical path through which the Italian anthropologist formulated his concept. Starting from Heidegger's idea of die Wohnenden ("the inhabitants"), De Martino builds the concept of territorialization, while also taking up the concept of bauen ("to build") theorized by the German philosopher, which derives from the ancient high-German buan ("to stay", "to remain") and which constitutes, going back to the genesis of Saxon languages, the concept bhu-beo, which is the same as ich bin ("I am"). De Martino also derived from Heidegger the other meaning of bauen linked to the idea of "constructing", understood in turn as "looking after". In this way, building not only identifies the concept of "construct" but also means, above all, to remain preserved from threats, thanks to the progressive expansion of knowledge that domesticates the unknown by making it known [6].

The analogy between CEDs, FCEG, and the cultural idea of Made in Italy product allows us to hypothesize, as shown in paragraph four of Fig. 2 that our participants' lack of full understanding of the first two categories of products. Their parallelism could be explained through the probability that they have not yet fully understood ethical jewellery to the same degree to which they currently understand the concept of Made in Italy products, due to a deep economic and historical embodiment. In this case, the analogy provides a frame wherein new insights and inferences are blended with known experiences. Through this process, the concept of ethical jewellery can gradually be made familiar and therefore begin to be easily understandable and incorporated into lived experience. In accord with Sanga, in this way, a cultural act becomes "natural" [52]. The mechanism of blending unfolds via the cultural appropriation of new concepts (e.g., Canadian diamonds) becoming understandable and eventually familiar through consumer perceptions of a new concept as similar to, or even an extension of, existing concepts (e.g., the Made in Italy brand). Thus, as we demonstrate through this study, consumers can perceive messages regarding new brands only within the boundaries of language and local culture. Once a new concept has effectively been assimilated into the familiar, both culturally and linguistically, the concept itself automatically becomes understandable; a workable frame has been provided, which in turn drives purchase decisions and, in the context of ethical jewellery, allows customers to behave ethically. Our considerations go even farther, proposing a new category that is not only theoretical but also practical and thus useful for marketing, which we define as "fair and ethically made in Italy". With this category, we can achieve a deeper understanding of how Italian consumers conceive, through their cultural background, the concepts of "fair" and "ethical" concerning jewellery and, in general, luxury goods. From an application point of view, this new category can also be a useful tool for understanding the perspective of consumers in Italy. Ultimately, the concept of "fair and ethically made in Italy" can be a vital component in creating ad hoc proposals for the Italian market aimed primarily at the consumption of quality or niche products, such as goods classified as ethical and sustainable.

References

1. Akrivou K, Di San Giorgio LT (2014) A dialogical conception of Habitus: allowing human freedom and restoring the social basis of learning. Front Hum Neurosci 8:1–4. https://doi.org/10.3389/fnhum.2014.00432
2. Arsel Z, Bean J (2013) Taste regimes and market-mediated practice. J Cons Res 39(5):899–917. https://doi.org/10.1086/666595
3. Banet-Weiser S (2012) Authentic: the politics of ambivalence in a brand culture. New York University Press, New York
4. Berger IE, Cunningham PH, Kozinets RV (1999) Consumer persuasion through cause-related advertising. In: Arnould E Scott L (eds) Advances in consumer research. Association for Consumer Research. University of Toronto, pp 491–497
5. Berardini SG (2015) Presenza e negazione. Ernesto De Martino tra filosofia, storia e religione. Edizioni ETS, Pisa
6. Berardini SG, Marrafa M (2016) Presenza e crisi della presenza tra filosofia e psicologia Consecutio Rerum. Rivista Critica della Postmodernità 1(1):93–112

7. Bernstein L (1992) Opting out of the legal system: extralegal contractual relations in the diamond industry. J Legal Stud 21(1):115–157. https://doi.org/10.1086/467902
8. Bourdieu P (1977) Outline of a theory of practice. Cambridge University Press, New York
9. Bourdieu P (1990) In other words: essays towards a reflexive sociology. Standford University Press, Redwood City, California
10. Bourdieu P (2000) Pascalian meditations. Standford University Press, Redwood City, California
11. Brønn PS, Vrioni AB (2001) Corporate social responsibility and cause-related marketing: an overview. Int J Advert 20(2):207–222. https://doi.org/10.1080/02650487.2001.11104887
12. Carrington MJ, Neville BA, Withwell GJ (2014) Lost in translation: exploring the ethical consumer intention-behavior gap. J Bus Res 67(1):2759–2767. https://doi.org/10.1016/j.jbusres.2012.09.022
13. Cervellon MC, Wernerfelt AS (2012) Knowledge sharing among green fashion communities online: lessons for the sustainable supply chain. J Fashion Mark Manage 16(2):176–192
14. Childs J (2008) Reforming small-scale mining in sub-Saharan Africa: political and ideological challenges to a Fair-Trade gold initiative. Resour Policy 33(4):203–209
15. Collier P, Hoeffler A (1998) On economic causes of civil war. Oxford Econ Pap 50(4): 563–573. https://doi.org/10.1093/oep/50.4.563
16. D'Angelo L (2019) Diamanti: Pratiche e stereotipi dell'estrazione mineraria in Sierra Leone. Meltemi, Milano
17. Davies NB, Krebs JR, West SA (2012) An introduction to behavioural ecology. Wiley-Blackwell, Oxford
18. De Boeck F (2001) Garimpeiro worlds: Digging, dying and "hunting" for diamonds in Angola. Rev Afr Political Econ 28(90):549–562. https://doi.org/10.1080/03056240108704565
19. De Martino E (1977/2008) La fine del mondo. Einaudi. Torino
20. De Theije M, Salman T (2018) Conflicts in marginal locations: Small-scale gold mining in the Amazon. In: Lahiri-Dutt K (ed) Between the Plough and the Pick. Informal, artisanal and small-scale mining the contemporary world. Australian National University Press, Acton, pp 261–274. https://doi.org/10.22459/bpp.03.2018
21. Duranti A (2001) Performance and encoding of agency in historical-natural languages. In: Henning K, Netherton N, Peterson LC (ed) Texas linguistic forum, vol 44. University of Texas Press, Austin, pp 266–287
22. Duranti A (2004) Agency in language. In: Duranti A (ed) A companion to linguistic anthropology. Blackwell, Malden MA, pp 451–473
23. Duranti A (2011) Ethnopragmathic and beyond: intentionality and agency across languages and cultures. In: Baraldi C, Borsari A, Carli A, Aurora CO (eds) Hybrids, differences, visions: on the study of culture. Davies Group, Aurora, pp 151–168
24. Echeverri P, Skålén P (2011) Co-creation and co-destruction: a practice-theory based study of interactive value formation. Mark Theory 11(3):351–373. https://doi.org/10.1177/1470593111408181
25. Evans-Lacko S, Henderson C, Thornicroft G (2012) Public knowledge, attitudes and behaviour regarding people with mental illness in England 2009-2012. Br J Pyschiatry 202(55):51–57. https://doi.org/10.1192/bjp.bp.112.112979
26. Fuentes JM, Martìnez-Jurado PJ (2014) Lean management, supply chain management and sustainability: a literature review. J Clean Prod 85(15):134–150. https://doi.org/10.1016/j.jclepro.2013.09.042
27. Giddens A (1979) Central problems in social theory: action, structure, and contradiction in social theory. In: Action, structure and contradittion in social analysis. University of California Press, Los Angeles
28. Giddens A (1984) The constitution of society: outline of the theory of structuration. In: Outline of the theory of structuration. University of California Press, Los Angeles
29. Ghilani JL (2012) DeBeers' "fighting diamonds": recruiting American consumers in world war II advertising. J Commun Inq 36(3):222–245. https://doi.org/10.1177/0196859912453320
30. Gupta S, Pirsch J (2006) The company-cause-customer fit decision in cause-related marketing. J Consum Mark 23(6):314–326. https://doi.org/10.1108/07363760610701850

31. Harvey J, Bolam H, Gregory D, Erdos G (2001) The effectiveness of training to change safety culture and attitudes within a highly regulated environment. Pers Rev 30(6):615–636. https://doi.org/10.1108/EUM0000000005976
32. Haufler V (2009) The kimberley process certification scheme: an innovation in global governance and conflict prevention. J Bus Ethics 89(1):403–416. https://doi.org/10.1007/s10551-010-0401-9
33. Hentschel T, Hruschka F, Priester M (2003) Artisanal and small-scale mining—challenges and opportunities. IED, London
34. Hilson G (2008) Fair trade gold: antecedents, prospects and challenges. Geoforum 39(1):386–400
35. Hilson G (2015) Fairtrade gold: antecedents, prospects and challenges. Geoforum 39(1):386–400
36. Jennings N (1999) Social and labor issues in small-scale mines. Report for discussion at the Tripartite Meeting on Social and Labor Issues in Small-scale Mines. ILO, Geneva
37. Joy A, Sherry JF (2003) Speaking of art as embodied imagination: a multisensory approach to understanding aesthetic experience. J Consum Res 30(2):259–282. https://doi.org/10.1086/376802
38. Joy A, Sherry JF, Deschenes J (2009) Conceptual blending in advertising. J Bus Res 62(1):39–49. https://doi.org/10.1016/j.jbusres.2007.11.015
39. Kapferer JN, Michaut-Denizeau A (2013) Is luxury compatible with sustainability? Luxury consumers' viewpoint. J Brand Manag 21(1):1–22. https://doi.org/10.1057/bm.2013.19
40. Keen M (1998) Vertical tax externalities in the theory of fiscal federalism. IMF Econ Rev 45:454–485. https://doi.org/10.2307/3867412
41. Kempf M (2017) Human development through fair gold mining? a case study of the fair mined certification. Dissertation, Lund University
42. Kjarsgaard BA, Levinson AA (2002) Diamonds in Canada. Gems Gemol 38(3):208–238
43. Ki CW, Kwon TH, Kim YW (2017) Luxury fashion consumers: comparing high and low guilt groups. In: Paper presented at the 2nd annual conference proceedings, international textile and apparel association (ITAA)
44. Kyba M, Giuliani G, Franziskakis F, Tockner K, Lacroix P (2019) Artisanal and small-scale mining sites in the democratic republic of the congo are not associated with nighttime light emissions. Multi Sci J 2(2):152–161. https://doi.org/10.3390/j2020013
45. Lujala P, Gleditsch NP, Gilmore E (2005) A diamond curse? civil war and a lootable resource. J Conflict Resolut 49(4):538–562. https://doi.org/10.1177/0022002705277548
46. Moraes C, Carrigan M, Bosangit C, Ferreira C, McGrath M (2017) Understanding ethical luxury consumption through practice theories: a study of fine jewellery purchases. J Bus Ethics 145(3):525–543. https://doi.org/10.1007/s10551-015-2893-9
47. Muskawa K (2014) Thematic lessons from the marikana miners' strike in South Africa: a corporate governance perspective. Mediterr J Soc Sci 5(2):63–67. https://doi.org/10.5901/mjss.2014v5n2p63
48. Okatei P, Cloutier C, Lehoux N (2017) Collaborative mechanisms for sustainable supply chain: a systematic literature review. In: Paper presented at CIRRETELT-2017–29, Université Laval, Québec, 14 June 2017
49. Reckwitz A (2002) Toward a theory of social practices: a development in culturalist theorizing. Eur J Soc Theory 5(2):243–263. https://doi.org/10.1177/13684310222225432
50. Rettie R, Burchell K, Riley D (2012) Normalising green behaviours: a new approach to sustainability marketing. J Mark Manag 28(3–4):420–444. https://doi.org/10.1080/0267257X.2012.658840
51. Røpke I (2009) Theories of practice — New inspiration for ecological economic studies on consumption. Ecol Econ 68(10):2490–2497. https://doi.org/10.1016/j.ecolecon.2009.05.015
52. Sanga G (1997) The domestication of speech: toward a theory of glottology. Europaea 3(1):65–120
53. Sasso G (2001) Ernesto De Martino fra religione e ilosoia. Bibliopolis, Napoli

54. Schatzki TR (1997) Practices and actions a wittgensteinian critique of bourdieu and giddens. Philos Soc Sci 27(3):283–308. https://doi.org/10.1177/004839319702700301
55. Schatzki TR (2001) Subject, body, place. Ann Am Assoc Geogr 91(4):698–702. https://doi.org/10.1111/0004-5608.00268
56. Schlosser K (2013) Regimes of ethical value? landscape, race, and representation in the canadian diamond industry. Antipode 45(1):161–179. https://doi.org/10.1111/j.1467-8330.2012.00996.x
57. Shove E, Walker G (2010) Governing transitions in the sustainability of everyday life. Res Policy 39(4):471–476. https://doi.org/10.1016/j.respol.2010.01.019
58. Sippl K (2015) Private and civil society governors of mercury pollution from artisanal and small-scale gold mining: a network analytic approach. Extr Ind Soc 2(2):198–208. https://doi.org/10.1016/j.exis.2015.01.008
59. Strahilevitz M, Myers JG (1998) Donations to charity as purchase incentives: How well they work may depend on what you are trying to sell. J Consum Res 24(4):434–446
60. Strahilevitz M (1999) The effects of product type and donation magnitude on willingness to pay more for a charity-linked brand. J Consum Psychol 8(3):215–241. https://doi.org/10.1207/s15327663jcp0803_02
61. Tylor C (1985) Philosophy and the human sciences. Cambridge University Press, Cambridge
62. Valerio G (2013) Making trouble: fighting for fair trade jewelry. Lion Books, Kidderminster
63. van Maanen M (2001) Phenomenology course handbook. Post-Workshop B. Ewha Women's University, Seoul
64. Bockstael Van (2014) The persistence of informality: perspectives on the future of artisanal mining in Liberia. Futures 62:10–20. https://doi.org/10.1016/j.futures.2014.02.004
65. Varadarajan PR, Menon A (1988) Cause-related marketing: a coalignment of marketing strategy and corporate philanthropy. J Mark 52(3):58–74. https://doi.org/10.1177/002224298805200306
66. Verbrugge B (2017) Towards a negotiated solution to conflicts between large-scale and small-scale miners? The Acupan contract mining project in the Philippines. Extr Ind Soc 4(2):352–360. https://doi.org/10.1016/j.exis.2016.10.011
67. Warde A (2005) Consumption and theories of practice. J Consum Culture 5(2):131–153. https://doi.org/10.1177/1469540505053090

Disrupting the Chain: The Luxury of Craftsmanship

Shaun Borstrock

Abstract The ways in which the luxury brand and designer craftsmen and women approach the design process could not be more different. The former is led by economics, vertical integration, mass production and global markets. Luxury brands manufacture millions of pieces of jewellery each season to ensure the fashion system is fully operational and profitable. The latter is led by an intrinsic understanding and interpretation of the world around them, their concern for the environment, the origin of their raw materials and the most appropriate integration of new technologies to enhance their product offer. Theo Fennell, Ute Decker and Mark Bloomfield are three London jewellers who approach design and making in two distinctively different ways. Fennell works predominantly in precious metals and stones, Bloomfield is a world leader in 3D printing and Decker works in sustainable materials. This chapter explores their shared understanding of materials and how they are used, the impact they have on the environment and a deliberate attempt to reduce waste, adopt sustainable practices and enhance the customer experience through hands-on engagement with their products.

Keywords Luxury · Sustainability · Supply chain · Jewellery · Circular economy · Fashion · Recycle · Innovation

This chapter explores the notions of luxury and how it, as a concept, is used to add value to jewellery. Despite or because of the material value and the impact of a 'label' or brand label on the attributed value of a product, consideration is also given to those designer-makers who through their own concerns for the environment are addressing issues around sustainability, ethics and social responsibility through their work. The contemporary luxury brands market is dominated by the luxury conglomerates, those companies that own the majority of the global luxury brand jewellery companies and other fashion-related luxury brands. These companies were once synonymous with design and innovation, craftsmanship, rarity and provenance

---

S. Borstrock (✉)
University of Hertfordshire, Hatfield, UK
e-mail: s.borstrock@herts.ac.uk

I. Coste-Maniere and M. Á. Gardetti (eds.), *Sustainable Luxury and Jewelry*, Environmental Footprints and Eco-design of Products and Processes,
https://doi.org/10.1007/978-981-16-2454-4_2

of place. The conundrum today is how we reconcile global jewellery brands and their production and distribution methods with those of the smaller designer-makers. What is often masked in sophisticated marketing campaigns is the reality of the process. Super slick advertising campaigns and visual narratives are produced with exceptionally high production values promoting all sorts of products in various ways. Not limited to physical published materials, luxury brands who have been slow adopters of the digital space are now pursuing social media channels with fervour. They are not then under the scrutiny of advertising standards authorities and can in effect say anything they want about their products—true or not (Figs. 1, 2 and 3).

Luxury continues to rouse the imagination of designers, makers, manufacturers, consumers and marketers. No one is without an opinion. What is clear, regardless of what anyone says whether an academic or from industry is that there is no clear definition of luxury. And this is the problem. How can almost anything be luxury? An often-used quote from Christopher Berry from his book, *The Idea of Luxury* [1], raises an important question which is even more relevant today. He states that 'luxury can without hesitation be tacked on to almost any article of merchandise from pizzas to handbags, from a fountain pen to a dressing gown and done so presumably to make it more desirable and the more likely to be bought' ([2]: 10). We have seen an explosion of goods onto the market that call themselves luxury. A finely crafted piece of jewellery by Theo Fennell is in the same 'luxury' category as a piece that is mass-produced by Van Cleef and Arpels or Cartier. This despite both companies creating their own categories of jewellery. At the one end, those that are mass-produced include the Alhambra collection from Van Cleef & Arpels and the LOVE collection at Cartier. At the other end both produce what they categorise as 'high jewellery', those pieces that command the highest prices. Cartier has the 'collection' pieces and Van Cleef & Arpels has the 'jewellery' collection that is less expensive and could be considered a diffusion line. The skill of the craftsman or woman is not considered in these categories. All three of these companies produce luxury jewellery. Then we have the designer-makers like Mark Bloomfield who combines technology and craft and Ute Decker also a designer-maker. Both are craftsmen, both are conscious of the impact of design on the environment and champion issues around sustainability. And like Theo Fennell, Bloomfield and Decker pursue the recycling agenda, albeit in different ways. Bloomfield suggests that:

> Luxury has become more widespread, accessible and popular in the last 20 years but it hasn't really changed. It remains excessive and surplus to one's need. But attitudes towards luxury are changing because of this. Investing in luxury 20 years ago as a means to exhibit your social standing or wealth was seen as aspirational, but now with luxury being so readily available it's become democratised and available to all. Anyone can indulge in a luxury chocolate bar and experience the dream prophesied by the marketer. What has changed is the realisation of the shallowness of that proposition. As you indulge you may question the activity and realise that it's out of alignment with the aspiration being sold by the luxury brands, particularly as concerns are raised around consumption, sustainability and what is really important? [3]

The power of persuasion is critical to ensure the consumption of any product. However, there is a stark contrast in spending power between those global brands and designer-makers. The stories told encompass anything from the rarity of the stone, the craftsmanship, origins of materials, sustainability strategies and the health and wellbeing of the workers. But dig deeper and more questions remain unanswered than answered. What is undeniable is that luxury as a concept is continuously changing but the manifestation of mass-produced fashion, in all its guises, as luxury is undoubtedly having an impact on how luxury is perceived. If one is to consider skill, materials, innovation, and craftsmanship as defining factors of luxury, a distinction needs to be made between what is and is not luxury.

1 Craftsmanship

> Proper craftsmanship jewellery, which is beautiful, thoughtful, sentimental, the sort of jewellery that you and I go to the museum and go oh my God, imagine who gave that. That really pulls at people's heart string, but it's not the one that's advertised, it's the other two ends that are advertised. Pandora at one end and Cartier at the other. The one in the middle can't afford to advertise, the craftsman thing, but what it needs is a sustained movement which I think is happening, of people talking about jewellery or silverware or whatever, a big arts and crafts movement [4]

Craftsmanship has long been associated with and attributed to luxury. The two exist in unity and no more so than with the creation of fine jewellery. It takes skill to work with precious metals and stones. Not limited to producing design work, craftsmen must also have an acute understanding of materials and how they react to varying temperatures, weight, material composition and physical manipulation. The same applies to precious and semi-precious stones. It is not necessarily the stone that defines the item produced but how the stone is used. Craftsmanship is about an intrinsic understanding of and relationship to one's materials. Theo Fennell, the London jeweller, has worked in all areas of jewellery production from bespoke to limited runs of gold, silver and semi-precious and precious stones. In my interview with him he said of craftsmanship:

> I would say the jewellery craftsmanship area, it's about the story of the artist, it's about what he or she is trying to put over. It's about the skill, it's about all those things, and nobody says, well hang on, how big is it, how much paint is in there, with jewellery in that middle bit, gold and stone and whatever are purely the paint. The rest is about the skill and about the concept and about the talismanic qualities [4]

This interpretation of craftsmanship, a comparison to the artist's materials and those of a jeweller, is the one that raises pertinent questions around how the consumer makes a purchasing decision. More often than not it's the name attributed to the product that justifies the sale of an artwork. One does not consider the materials. The craftsmanship is dismissed in favour of the 'brand'. With jewellery it is arguably the brand that determines the purchase. Tiffany & Co or Van Cleef & Arpels, two of the most prominent jewellery companies in the world, are defined by their brand image rather than the quality of the product or materials used.

Fig. 1 Theo Fennell—stonehenge ring

Typically, a craftsman or woman has honed his or her skills over many years through apprenticeships, learning on the job and constant practice; in effect, becoming one with their tools and materials. The process starts with an idea that develops into a design which is realised, a process conceived by the craftsman or woman. This process differs immensely from that of a 'design factory', those manufactures serving a global consumer base with multiple products sold in multiple stores. Comparisons are drawn up to unpick the different approaches taken to produce jewellery in different settings—a factory and a workshop. What is undeniable is that there is indeed a difference. Craft is the constant refinement of an activity that manipulates material in a considered way. As Bloomfield suggests:

> craftsmanship is the skill accumulated through responding to materials with intent through the use of tools that enable the refinement of an individual's craft. It feels like an obsessive process, one that draws the craftsperson into a personal dialogue with the activity as they search for a result or answer that wasn't evident at the beginning of the process. It can be likened to a voyage of discovery [3]

If for the purposes of this argument one is to differentiate the craftsman or woman from the factory worker, clear boundaries are set. The products could then be categorised as luxury and luxury branded. Creating this distinction allows us then to articulate the difference. Luxury jewellery is crafted by a craftsman or woman described by Richard Sennett in *The Craftsman* [5] as someone renowned for their 'skill, commitment, and judgment' focused on 'the intimate connection between hand and head'. He goes on to make two 'contentious arguments: first that skills, even the most abstract, begin as bodily practices; second, that technical understanding develops through the powers of imagination'. A number of authors extol the values of the touch of the hand as a defining factor to describe something as luxury. And the touch of hand in this instance is inextricably linked to tacit knowledge, where the maker is at one with their materials. David Pye, the author of *The Nature and Art of Workmanship*, stated that 'workmanship using any kind of technique or apparatus, in which the quality of the result is not predetermined, but depends on the judgment, dexterity, and care which the maker exercises as he works' [6]. This interpretation of workmanship is, it could be argued, true of craftsmanship. The ability to undertake a task where the end result is dependent on the skill of the maker illustrates a commitment to perfection, or as close to perfection as they may get. It is akin to being able to draw a full circle in one movement. This differs from what Pye considers to be workmanship of certainty, where the outcome of the end product is predetermined. This applies to those products that are mass-produced, like a chain, or a product made through a process of injection moulding.

Ute Decker is a London-based jeweller who was one of the first to publicly express her concerns about questionable labour practices and the impact that mining has on the environment. She only uses Fairtrade Gold, and is one of the few jewellers who do so. Decker describes her work as an expression of freedom, one where the final outcome is the result of practice and instinctive response to the Fairtrade gold she uses.

> It's an intuitive relationship I have with the shape. I just keep making it in brass so when I come to make it in the precious metal, I [have] what the Japanese call an empty mind. My brain is no longer involved, I'm not thinking of anything because my hands know what to do [7].

The reference to the Japanese empty mind, or Mushin,[1] very much echoes Sennett's suggestion relating craftsmanship to bodily practices and the power of the imagination. This approach is in stark contrast to a factory setting. 'Within a factory setting where the worker is assigned the arduous job of repeating their task over and over again, doing the same thing day in and day out there is little requirement for them to undertake 'bodily practices', but more importantly the powers of the imagination. A craftsman or woman is defined by his or her skill, 'the good craftsman, moreover, uses solutions to uncover new territories; problem solving and problem finding are intimately related in his or her mind' [5]. Uncovering new territories defines the craftsman. The ability and desire to challenge oneself in order to discover new ways of doing something is a challenge that all craftsmen or women set for themselves. They want to innovate. If one considers two very different makers, Louis Vuitton and Carl Faberge, both were craftsmen albeit in different fields. What they had in common is that they were both innovators. And both focused on their own areas of expertise. There was no diversification of their craft. These traits define the craftsman or woman, they represent the inherent differences between the work of a craftsman or woman and one who works on a production line, even if the goods that both produce are presented as luxury. Reverting back to Pye, his concern and indeed a concern of many craftsman is that 'The danger is not that the workmanship of risk

[1] 'Mushin is the essence of Zen and Japanese martial arts. It is a state of mind where the mind is not fixed on or occupied by any thought or emotion and is thus connected to the Cosmos. In Zen on in your daily life, if the impulse is expressed as conscious thought, it is not Zen. This pure state of mind, of pure mental clarity, is produced by the absence of the ego or limited self. A Mushin mind is not an empty mind like an empty shell; on the contrary, it is a mindfully present, aware, and free.' [8].

will die out altogether but rather that, from want of theory, and lack of standards, its possibilities will be neglected and infer forms of it will be taken for granted and accepted' [6]. And this is a precarious position to be in, and one that looms ever closer as mass production defined as luxury negates the skill of the craftsman or woman in the public domain. In addition, one must consider the impact of technology on the manufacturing process. When Faberge was producing his jewellery and Imperial Easter eggs, they were all produced by hand. This process was not dissimilar to that of Van Cleef & Arpels during the same time. But times have changed, and Van Cleef & Arpels now mass produce their most successful jewellery collection, the Alhambra, to keep up with demand. That is not to say that no handwork is employed as part of the stone setting and polishing processes. It is all about scale and the ability to meet the demand; demand that they, Van Cleef & Arpels, have created. According to Fennell, since they were acquired by Richemont

> they did it up, and gradually I started to see these full page advertisements for Alhambra, the 50th Anniversary, the new Alhambra, just exactly the same stuff, but I mean full page [in the] Telegraph, full page everywhere, and you know what, I don't know where it is now, but three years ago it was the bestselling jewellery collection in the world. So, like nothing else, that showed me the power of that remorseless advertising [4].

The power yielded by the conglomerate illustrates an incessant desire to create desire and to encourage mass consumption. The principle is simple. Use the high-end product that is crafted by master craftsman or woman to sell the products that are easily mass-produced but retain the value of the more, unattainable product offer. This is a common route to market for all luxury brands. It is the heritage built on many years of expertise, before diversification, that adds value to those products that are readily available. Although the manufacturing processes have changed, enabling faster production and distribution, it is the aura behind the brand that disregards the production method in favour of a historical narrative that maintains the brand value.

As luxury and the definitions thereof change, mostly in favour of the creation of a myth founded on heritage and true craftsmanship, it is important to identify the deceptive adoption of heritage to create value where it may not exist. The approach undermines the skill of the craftsman or woman and the inherent authenticity of their craft.

2 The Supply Chain—From Corrupt to Care

Consider the supply chain where at one instance it fulfils the demand and at the other it creates it. The three jewellers I have interviewed for this chapter are very clear about their design process. They are designer-makers, craftsmen and women. But as I have already said their approaches to design and make are found based on different principles and approaches. But what they do have in common, which is something that is at the heart of nearly all craftsmen and women, is the desire to continue to produce quality pieces. At the same time, they are fully aware of their actions and the impact they have on the world around them. Their philosophical approach to design and make is worlds away from the luxury brands such as Cartier, Van Cleef & Arpels and Tiffany & Co (although this was not always the case). Tiffany & Co, which has the biggest retail presence out of the three, produce more items of silver jewellery for the mass market. It is those low-ticket items that are more accessible than the fine jewellery items. Mark Bloomfield incorporates both technology (CAD and 3D printing) and traditional making skills as part of his approach to design. Fennell is a traditionalist at heart, he works with pencil and paper, and Decker takes an intuitive approach to realise her designs. All three have an insatiable appetite for longevity, legacy and raising concerns about, and protecting, the world in which they live.

When one attributes technical terms to define mass production processes it is difficult to reconcile them with independent producers, but not difficult to do the same for global brands. The supply chain has, over the past 100 years, become increasingly sophisticated and is constantly evolving. Historically, it was predominantly a regional process, limited in some instances to three or four transactions. The farmer harvested the wheat, sent it to the mill to be ground to flour and the miller sent it to the baker to produce the bread. Today, the supply chain is, of course, much more sophisticated. And as Industry 4.0 takes hold, the process will be much more reliant on automation, data farming and better integration between the physical and digital worlds. Production will be defined by smart processes and data will determine the steps of production. This ultra-mechanised process will eventually lead to autonomous decision-making with those decisions processed in real time. This will place further emphasis on vertical integration and its capacity to be even more reliable. This adds to the power yielded by the luxury brand conglomerates as they continue their pursuit to acquire more key contributors to ensure expediency within their own supply chains.

> One of the biggest influences has been the explosion of manufacturing in Asia, with China, Japan and Korea becoming major suppliers and exporters of goods. At the same time, AI and machine learning combines with predictive and prescriptive analytics to provide better forecasting, enhanced order management and more. What's more, the supply chain is evolving toward a more data-driven, network-driven and collaborative supply chain ecosystem that drives real value and growth for all participants [9].

With the implementation of numerous technological advances and the drive to reduce costs, big corporations are looking at new ways to reduce waste and at the same time increase profits. Some smaller companies are concerned about the enormous amount of waste produced and are actively looking to reduce their carbon footprint, take action against untoward working practices and hold companies accountable for their actions. These concerns are emergent and continue to gain traction but are not a common concern amongst the global jewellery trade. Evidence of this is the constant supply of jewellery often shown at trade shows around the world. As Fennell says

> every time I go to Basel or another trade fair, every time I get sent an auction catalogue from Christie's or Bonham's, I think how much jewellery is there in the world? I remember the first time walking into a jewellery trade show and the hall was the size of the Albert Hall, full of jewellery and watches. 95% of all the jewellery that's made, or something like that, is non branded. We think of Cartier, Tiffany, we think those are the jewellers and then there

> are few county jewellers and a few high street jewellers, but they are only 5%. All the other stuff is the stuff that's made in India, China, it just goes out, it's just stuff [4].

Within the world of jewellery there are many areas of contention—the origins of precious and semi-precious metals and stones and the labour conditions of those who are on the front line of excavation and mass production. However, mass production fuels mass consumption.

> Consumption is based on waste as consumers are constantly encouraged to buy more without the means to successfully recycle the things they already own. Therefore, most unwanted product ends up in landfill. The buy and discard cycle is becoming a global issue as we're made aware of how our consuming behaviour is impacting our natural resources. Also supply chains are so efficient and effective that 'more' is always available [3].

It is the constant stream of goods that increases waste and adds to landfill sites around the globe. Typically, a production line ensures the expediency of supply no matter what is in the chain. These complex chains of sourcing, manufacture, distribution and retail all consume huge natural resources. The destruction of land to make way for mines and extract the raw material, the fossil fuels burnt to create power, ensure the efficient transport of goods to wherever it may be that they need to go, all these procedures contribute to a continuous chain of supply and demand. The creating of raw and finished materials for packaging and of course the considerable resources consumed to power create the promotion of the products and services. These typically destructive processes are only different as a result of how they are conducted; one may be a factory producing goods, the other a team of creatives flying to far-flung destinations to do a photographic shoot for an advertisement. They all, however, function under the division of labour to ensure efficiency through the allocation of tasks assigned to individual works to ensure maximum efficiency. In his book Mass Customisation, Joseph Pine [10] considers in detail, the impact of mass production on consumption and vice versa. He states that 'in mass production, companies increased output not only by adding inputs but by increasing the throughput of machines and the productivity of the workers so that fewer workers were needed per unit produced'. He goes on to say that 'this greatly increased fixed costs and the capital-to-labour (machine-to-worker) ratio, but also greatly lowered costs of each unit' ([10], p. 16). And as the supply chain becomes more sophisticated and dictated to by technological advancement, corporate power becomes stronger, more demanding and more focused on improvements that ultimately lead to increased profits. With jewellery, especially those that are made in multiples or mass-produced like, for example, the Alhambra collection from Van Cleef & Arpels, or the LOVE collection by Cartier, 'the product must be as standardised as possible and the manufacturing process broken down into small specified tasks. Workers and machinery must be highly specialised to drive down the time and expense of each individual task' [10]. There may be some hand setting and polishing, but the heavy-duty tasks are undertaken by the machine. Ultimately, the touch of hand is limited as is the element of risk. These are two crucial components that contribute to the making of a luxury product. With Van Cleef & Arpels and Cartier, for example, one is not exposed to the makers and they work anonymously behind the scenes. The design process is hidden as is that of the manufacture. Adding to this, the acquisition of suppliers to a consolidated supply chain that in principle has the capacity to service all brands and the supply chain becomes increasingly opaque. One is then left in a quandary. What then becomes problematic is when one attempts to delve into issues of concern around origin of materials, distribution channels and labour practices within a global luxury brand. No information is forthcoming, and the deeper you dig the less you find out.

There is a distinct lack of clarity around the jewellery luxury brand supply chain. Decker suggests that there is a lot 'of ugliness in gold and silver mining and also financial civil corruption. It's all very ugly. She goes on to say that there is no traceability in gold as it comes from everywhere in the world and is usually melted in refineries in Dubai and then shipped to its destination' [7]. Fennell echoes this thought by saying that:

> my belief is going forward people are going to be very concerned about where the gold comes from and where the stones come from, and I think the origin of the materials is becoming far more important than it was. And in a slightly ironic way, people believe that recycled jewellery is better than new jewellery, unless you can genuinely prove where it came from. The irony of that is the people who mined the early diamonds, the early gold, you can almost guarantee that they diamond mined in India or Africa before 1920 let's say, was the result of really appalling conditions and of slavery [4].

Although Fennell refers to the appalling conditions of the 1920s, there is no doubt that those practices still exist today. Blood gold and blood diamonds, where both are mined under the most horrific circumstances, elicit profits at the expense of the labourers. They are practices that continue to be brought to light by amongst others, smaller designer-makers like Decker and Lebrusan Studios. They and others are attempting to end this practice by increasing pressure on the mines 'to guarantee the provenance of gold, the group that oversees the world's largest bullion market is seeking new ways to securely track each step from the mine to the jewellery store' [11]. In theory, this approach is admirable but in practice dubious. When trying to find information about the ethical and sustainable practices from any of the luxury brand websites, there is a distinct lack of transparency. This is in contrast to Decker who can track her gold to the mine. Of the process she explains the following:

> I use Fairtrade gold which is more expensive, and you have to be a Fairtrade license holder. But there is also a team because it's fully traceable, so you need to be registered in the system. The mine has a Fairtrade number. The importer has a Fairtrade number. The refiner has a Fairtrade number and I have a Fairtrade number [7].

Fig. 2 Ute Decker—rolling waves in moonlight

The supply chain that Decker describes is one of accountability. However, they can be used to achieve very different aims. Bloomfield says that 'supply chains are an effective way to realise efficiency in the production process, but they can also be used to mask or hide various activity that a company may not want to reveal about its business. Sometimes a part of a company's supply chain may be firewalled behind a non-disclosure agreement for example' [3]. These opaque systems put in place by the luxury brands are in essence a distraction from the realities of getting a product to market. These processes are further masked by the relentless advertising campaigns employed to promote their products and the virtues thereof. It would, under these circumstances, be nonsensical to promote the true origins of a product and how the materials were sourced and then processed for ultimate consumption.

On the other hand, as is apparent with Bloomfield and Decker, full transparency of their supply chain is refreshing as it provides a unique and vitally important, insight into the world of, in this instance, the supply chain and its relationship to, and impact on craftsmanship. Jewellery exists in many guises, from plastic to base metals and silver to gold and other precious metals and stones. But it has, and in some instances still does, encompass luxury in an almost unique way. Craftsmanship plays a considerable part in defining jewellery as a luxury. And through marketing, diamonds and other precious stones and metals have also come to define jewellery as luxury. This latter point is debatable as will be discussed. Fennell puts it that there are tiers of jewellery where at the top end the maker defines it through craftsmanship. Bloomfield concurs with Fennell who says:

> traditional jewellery had a relationship with wealth and power through the use of precious materials. A gold crown for example on a Head of State relates the material, i.e. gold, to high standing individuals that would use that material as a symbol of wealth and power. Often employing highly skilled craftsman to create extraordinary pieces that would be renowned the world over to elevate that individuals social standing. The person that has the skill to manipulate the material to that degree also establishes links in a supply chain that has not really changed. Contemporary precious jewellery not only carries the intrinsic value based on a moment, or emotional response, becoming engaged for example, but it also carries that association to the preciousness of the material [3].

However, it could be argued that luxury branded fashion clothing and luxury branded jewellery, in contemporary terms, have become somewhat complicated to define. Both have tiers or categories that are defined through levels of aspiration and disposable income. At the bottom end, one could buy a bottle of fragrance, and at the top end, one could commission a bespoke fragrance. At the high end in fashion exists haute couture and with jewellery there are bespoke or commissioned pieces. Both command a price, service, the finest materials, the craftsmanship and attention to detail—and an element of rarity and scarcity. Fennell clarifies the jewellery system as:

> three main groupings, where one is power jewellery, and that is I own you by giving you this £2 million ring. I'm a king therefore I'm wearing a crown, he must be rich. Look at the stone his wife is wearing or look at the watch he's wearing, and that's happened forever. You know the king of Versailles had a great big crown or whatever, it's the most obvious end of jewellery. The far end is just self-decoration, and that's just something that everybody had

> done forever to make themselves feel better, whether it's a piece of coral or a piece of gold or whatever it is. In the middle, is something that is neither of those things, which is proper jewellery. Proper craftsmanship jewellery, which is beautiful, thoughtful and sentimental [4].

All sectors of the luxury trade are beholden to the supply chain. The issue is how it can be disrupted to ensure a level of accountability and indeed responsibility is at the forefront of design thinking and realisation. As discussed, there are significant areas of concern relating to how materials are mined, plastics created, and waste generated. Within the luxury branded goods and services environment all sectors also suffer from over production, over consumption, fabricated mythology and the trends often dictated by the big three conglomerates. There is an incessant appetite to continually grow their businesses, encourage consumption and further streamline their production through enhanced vertical integration. All these examples influence the functionality of the supply chain and its irrefutable impact on our consuming habits.

3 Diamonds Are Forever, or Are They?

Consumers have long been encouraged to part with their money through marketing strategies that glorify a product. Diamonds are no exception.

> Americans exchange diamond rings as part of the engagement process, because in 1938 De Beers decided that they would like us to. Prior to a stunningly successful marketing campaign 1938, Americans occasionally exchanged engagement rings, but wasn't a pervasive occurrence. Not only is the demand for diamonds a marketing invention, but diamonds aren't actually that rare. Only by carefully restricting the supply has De Beers kept the price of a diamond high [12].

De Beers has long been the name associated with diamonds, unscrupulous working conditions for the miners, and flooding the market with diamonds. They are no longer the largest diamond mining company in the world and they no longer hold a monopoly, 'in 2018, they mined 35.3 million-carats of diamonds' [13]. That accolade belongs to the Russian company Alrosa having turned out 36.7 million carats of the total of 43 million diamonds produced in Russia in 2018. Canada, the world's third-largest diamond mining nation, produced 23 million carats in 2018 [14]. If one is to consider this yield as one year's output, 101 million carats produced, then one would agree with Dhar's suggestion that diamonds are not in fact rare. Despite their physical sparkle postproduction, diamonds have a history mired in strife. Diamond mining has been beset with problems most of which have been raised publicly by human rights activists, class action lawsuits and the miners themselves. These actions have forced change with the introduction of some measures like the Kimberley process, to protect the workforces and reduce the flow of blood diamonds onto the market. Certified diamonds, and in some instances some kinds of gold, are being seen as 'guilt free' and heavily promoted by the retailers as a sustainable and responsible purchasing solution. But as already stated there is still some way to go to be 100 percent certain of the origins of a stone or metal. Despite all this, and if one were to then consider the prices of diamonds at retail, it is astounding that the retailers are able to justify their prices and that luxury branded jewellery, in the main, maintains its value. According to Bloomfield:

> All materials have a value, some more than others and even plentiful materials can be transformed into something of great value. Precious jewellery has a history of being made from materials that have an apparent high market value, gold and diamonds for example, but is sold based for its intrinsic value. Often the value of precious jewellery is maintained through the belief in the value of the materials, their rarity, and the work involved in its realisation. But a more important value proposition is how it makes the customer feel. Customers are beginning to realise that value can come from a variety of sources, precious or otherwise. Particularly as the value of precious materials are questioned, diamonds for example achieved their high value through restrictive control of the market, even though there was and still is a plentiful supply [3].

Jewellery has a particular place in the market despite the substantial damage mining of any kind does to the environment. Up till now, the customer has not really considered this as an issue. And most people still do not think about the damage caused to the environment in the production of a piece of jewellery. The response by the corporations, including De Beers, Van Cleef & Arpels, Cartier and Tiffany & Co, to environmental concerns is that 'Not only do we have this direct traceability but we can demonstrate the positive impact your diamond has had along its incredible journey' [15]. And 'At Tiffany, we directly source the majority of our diamonds from known mines and sources' [16]. Van Cleef & Arpels state that they 'continuously seek[s] to improve its environmental management in order to reduce its impact, and focus on finding innovative solutions, that promote a circular economy'. These companies justify their actions, as do many more, by suggesting that it is acceptable to mine for diamonds and gold (which is a much more toxic process when extracting the ore from the ground as a result of the use of the noxious chemicals) if they support the workforce, plant trees and enhance the provision of learning. It is apparent that the jewellery world tends to exaggerate the benefits of their products. There is never a mention of the point of origin of their products. One no longer sees an advert for a gold and diamond necklace that extols the virtues of the origins of the stone or material. Images of the mines are not used in the advertising regardless of where the stones come from. Nor are the consumers confronted with advertising that promotes the apparent positive impact mining has on communities and the environment. It goes without saying that with an increase in production the purchasing power of companies is stronger, more raw materials are produced and the labour costs go down. Increased productivity results in a reduction of overall manufacturing costs. It is therefore safe to say that despite reduced manufacturing costs the savings made by the manufacturer are not passed on to the customer. The opposite is in fact the case.

Savings makes business sense. With gold and diamonds the classification of both creates differentiation. Diamonds and gold are both categorised in carats. With diamonds the size, colour and clarity of the stone is often what contributes to its price.

Gold classification ranges from 9ct to 24ct, with the latter being the most expensive. With diamonds, the bigger and clearer are the better ones. Diamonds, however, can be 'clarity enhanced'. 'Advanced technology nowadays has enabled the jewellery industry to improve the visual appearance of lower grade diamonds by the process of laser drilling or fracture filling' [17]. Enhancing the brilliance of the stone adds value. If a stone is unheated and naturally brilliant in colour it would make sense for it to be described as rare and for its price to be elevated. If, however, it has been treated to enhance the colour the opposite would, or should be, the case. Issues arise when a stone has its value based on what it would have been had it been naturally brilliant. This is a common practice and the information is not forthcoming from the jeweller. On searching any of the luxury brand jewellery websites there is no information about either the origin of the stone or whether it has been clarity enhanced.

> Some retail jewellers don't disclose the fact the diamond has been fracture filled or laser drilled. Jewellers can simply write the abbreviation "CE" for clarity enhanced on the paperwork, and an unsuspecting consumer would never know what that means. For those customers who wish to purchase diamonds as investments, treated diamonds cannot and will not appreciate in value or hold value as would an untreated diamond [17].

Technology plays a vital role in altering the colour of a stone to make it that much more desirable. Changing the colour of a stone from a yellowish hue to one that is brilliant in its clarity does one of two things. Firstly, it makes it more appealing to the customer; secondly, it adopts the attributes of a naturally occurring brilliance found in a stone where the brilliance has happened naturally in the ground. It may be enough to want a brilliant, sparkling diamond where the value may not increase, but the supplier must be held accountable to ensure the customer is aware of the pitfalls of buying a clarity-enhanced stone. Technology is also being used to grow diamonds in a laboratory:

> The only thing that makes a lab-created diamond different from a natural diamond is its origin. A lab-created diamond is "grown" inside a lab using cutting-edge technology that replicates the natural diamond growing process. The result is a man-made diamond that is chemically, physically, and optically the same as those grown beneath the Earth's surface [18].

Increasingly lab-grown diamonds are seen to be the favoured choice of generation Z and Millennials as the environmental impact is reported to be minimal. 'A 2018 study found that 70% of millennials said they would consider buying a lab-grown diamond' [19]. As the younger generation of consumers become increasingly concerned with the impact of our actions on the environment their decisions also start to impact the behaviour of luxury brands. De Beers, the first to take direct action have started to produce their own lab-grown diamonds. What is interesting is that they do not retail any of these stones on their main fashion store website or in their physical stores. They are produced by Element Six a subsidiary of De Beers who 'has developed to become a global leader in synthetic diamond and supermaterials manufacturing' [20]. These laboratory-produced stones are supplied to Lightbox jewellery who produces the products. Once again one finds oneself in territories that lack transparency. What is clear is that the retail environments of the jewellery brands owned by the luxury brand conglomerates do not, at the moment, sell jewellery set with lab-grown diamonds.

Three conglomerates control the global luxury goods jewellery market. LVMH, Kering and Richemont own between them the following jewellery brands:

LVMH	Kering	Richemont
Bulgari	Boucheron	Buccellati
Chaumet	Pomellato	Cartier
Fred	Qeelin	Van Cleef & Arpels
Tiffany	DoDo	

What is striking is that these 11 brands are probably the most recognised in the world. They are certainly the most profitable with hundreds of outlets between them. They may be standalone stores, concessions, in airport duty-free malls and/or discount designer villages. They all promote their wares as luxury and unashamedly promote craftsmanship from a very 'unique' angle. Their selective disclosure of 'craftsmanship' promoted in their advertisements show hands polishing gold and cutting and setting stones. This fabricated reality bears no resemblance to the true adoption and implementation of craftsmanship employed by those companies less likely to mass produce products for a network of global retailers. The design process differs from company to company. But typically, and as an example, the designer starts with an idea that they then illustrate. They consider the three-dimensional qualities of the item, the materials that may be used and then start to realise the item in 3D. This could be through modelling with CAD, physical modelling with clay or 3D printing a model. These initial steps are the precursor to the production. Mass production may take a different turn where a mould is produced that would be used in the mechanised production process. Smaller designer-makers are more encouraging of the customer to engage with the process. This differs to the bigger companies whose processes are closely guarded and not disclosed other than through high production films that attempt to glorify the process.

4 Change—Green or Greenwashing

The process of design differs between sectors of the market. Jewellers like Bloomfield, Decker and Fennell, all craftsmen and women in their own right, approach their 'subject' very differently. Bloomfield uses a combination of traditional tools and technology.

> Technology is just another tool and I aim to use it with the same obsessive nature that I would when using a hammer or a file. It's important for the technology you use to be appropriate or you make it appropriate through the skill developed as it's used. Technology can be just as revealing, I use 3D modeling software for example to create a shape, I have the skills to achieve the shape I want but equally I'm able to use the technology in an

Fig. 3 Mark Bloomfield—orbital one

exploratory innovative way to discover shapes that I may not have considered. You can approach technology from a position of craftsmanship, to get the best out of the process or reveal something new that enhances that process or makes it innovative [3].

Fennell himself uses pencil and paper although his studio team does use CAD in the studio to render and virtually realise designs for their clients. They also use CAD and 3D printing to create templates when designing complex pieces. This avoids expensive tooling costs when perfecting the spout on a jug for example. Decker's approach is much more intuitive as she does not 'design' on paper at all and relies solely on her intuition. She refers to her manipulation of gold akin to working with ribbon. The one thing that brings them together is their belief that jewellery should not be disposable. It should have real meaning and legacy. And there should be a connection with the maker, something that further enhances the experience of owning a piece of their jewellery. Their approach addresses issues around the supply chain differently. They raise concerns for the environment, incessant consumption and unscrupulous labour practices. Each of them employs different methods that address the circular model to keep their work in circulation and not be disposed of. Bloomfield says of his design process:

I design and create products that can be reused and recycled. All the jewellery I make is based on a modular design ethos that allows it to be updated, changed, customised and reused. I use 3D printing as the manufacturing technology that allows me to make just one item or 1000"s that are locally manufactured. Customers are encouraged to play with the modular system to create one of a kind items and adapt items they already have rather than discarding them. I'm also making products from other recycled materials, plastic and paper for example. I use design and craftsmanship to increase the value of waste materials, and try to create a more meaningful story at the time of purchase so customers have a stronger connection to it [3].

It is evident that his relationships with his customers are important. And Bloomfield like Decker and Fennell value those relationships. There is a striking contrast between how they work and how the luxury brands work. With Cartier, Van Cleef & Arpels and Tiffany & Co, for example, one does not, typically, have the opportunity to work with a designer or the person who makes the product one buys. It is up to the sales associate to sell the product. And where traceability is concerned the sales associate would have the information about the origins of the stone or metal from Head Office. As one would expect, the response would be heavily scripted. For diamonds, for example, one would be told that they are traceable and not blood diamonds. However, there is little to no information, as Decker has attested, to the origins of the metals. Most jewellers sign up to a number of global entities such as the RJC (Responsible Jewellery Council) whose 'Code of Practice is the only independently verified sustainability standard for the global jewellery and watch industry covering the entire supply chain, from mine to retail' [21]. What is telling is the language used to describe their social responsibility agendas. A common approach by companies is to use language that is fluid and does not tie them to anything in particular. According to the Van Cleef & Arpels Corporate Social Responsibility Policy they 'strive to source its raw materials responsibly' and 'the Maison also expects that its suppliers communicate and encourage their own contractors and vendors to incorporate these principles into their business policies and practices' [22]. Cartier follows a similar trajectory stating that 'We will exercise the greatest vigilance with respect to our supply chain, especially regarding diamond purchasing' [23]. Their policies and ambitions are to be applauded but as with those luxury brands who have signed up to initiatives such as the RJC, it is incredibly difficult to find the information one needs to make an informed decision about their methods of track and trace and whether or not they are reliable:

> Companies need to take responsibility for the products they sell rather than terminating the relationship with the customer once the sale has been made. Manufacturers have all the insight into an understanding of the products they make and are in the best position to reuse and recycle that material once it's no longer needed by the customer. If they were to take greater responsibility then they would also have more consideration for the design of their products in the beginning, to make the recycling process easier in the end [3].

Companies could come clean; they could easily disclose the true origins of their raw materials and how and where their products are made. Blockchain management or track and trace are sophisticated tools that 'provide[s] end-to-end visibility into multi-enterprise supply chain networks, enabling asset owners to track and trace things of value to achieve faster results and establish trust between participating trading partners' [24]. One can see why, for obvious reasons, the food chain would employ such erudite systems.

Track and trace in the food supply chain have become common practice with complex monitoring systems in place to satisfy the needs of both the suppliers and increasingly demanding consumers. Evidence of this would be on the packaging of a piece of meat in a supermarket that has the name of the farmer on it. The agricultural supply chain is incredibly complex due to many variations of, for example, the same type of fruit and vegetables grown. It is possible to track and trace a single vegetable from germination right the way up the supply chain to the end-user. A phenomenal amount of data exist that provide information relating to the water yield, soil and weather conditions, nutritional information, flavour and colour and so on. Despite the following quote being attributed to an agricultural supply chain, it is also wholly applicable to luxury goods. Schuster suggests that 'Consumers tend to depend on branding for making quality and purchase decisions rather than a comprehensive understanding of the origin of specific food products' [25]. Branding and branding strategies play a major part in persuading customers to purchase a product or service.

It is exacerbated by the global marketing campaigns to promote products and services with the ultimate goal of achieving customer loyalty and repeat purchasing.

The issue here is that very seldom is the luxury brand industry held accountable for their actions. For example, in 2018, Burberry as reported by the BBC and unlike many other luxury brands admitted to burning stock that did not sell. The BBC reported that they 'destroyed unsold clothes, accessories and perfume worth £28.6 m… to protect its brand. It takes the total value of goods it has destroyed over the past five years to more than £90 m' (Burberry burns bags, clothes and perfume worth millions, 2018). After much criticism the company released a statement as was reported in the Guardian newspaper that they were to 'end its practice of burning unsold clothes, bags and perfume and will also stop using real fur after criticism from environmental campaigners' [26]. This wasteful activity is not confined to Burberry but is a common practice across the luxury brands industry. Despite their public support of the sustainable and ethical practice, they flout the causes they promote. This greenwashing is an unfortunate but common practice in the world of luxury.

However, unlike the food industry, luxury brands are not held accountable for poor quality products, poor working conditions and destructive practices that contribute to the production of their products. Saying that, there is clearly an emergent consumer group looking to better understand the origins of the products they buy. LVMH, Kering and Richemont all have social responsibility agendas but, as already discussed, they do nothing other than pay lip service to those agendas. This raises a number of questions as to why luxury brands continue to conceal the true origins of their materials, the composition thereof, their distribution channels and labour practices. With the rise in the consumption of organic goods, the use by some fashion companies of recycled plastics (Adidas, Stella McCartney), cashmere (All Saints) and low water yield denim (Warp + Weft), it is possible to provide some concrete information about the sustainable practice. To appease their customers, luxury brands make claims that they are indeed protecting the environment through sustainable practice but this is mostly greenwashing.

5 Conclusion

The focus of this article is geared towards the craftsman or woman and addresses industry practices. The jewellery companies mentioned all attest to protecting the environment, supporting the workforce, planting trees etc. But the devastation they cause is in effect unnecessary. It is in most cases irreversible. There are no environmental benefits to mining. In the case of the diamond, they can and are being manufactured. In short, the conglomerates go to extraordinary lengths to protest their innocence under the guise of sustainability when the opposite is in fact the case. But there are solutions. As Fennell says:

> I'm a great believer, and everybody who has a shoe box or whatever it is, full of old jewellery, should take it out, have it looked at, and it will come out in a lump of gold and some stones, and have something made by a craftsman. Certainly, the big brands are never going to do

that. It doesn't fit in in any way with their business model, but I think it's something that could really be the start of keeping the sentiment [4].

This idea, the implication, that there is so much unworn jewellery out there, much of which would benefit from remodelling is one solution to reimagining a sustainable future. This, a visionary path to reducing wasteful and environmentally unsound practice, would not only give life to unworn pieces of jewellery but also generate an unlimited supply of raw materials. Innovation does not rely upon a stream of new material but also encourages the reuse of old material. As Fennell suggests 'I think you really can change people's minds and I think when you do change people's minds about sustainability, about how things re-made, about getting involved in the process. Jewellery is the one that you really can remake, because the raw materials will always remain' [4]. In fact, this approach is not limited to jewellery it can be applied to clothing and accessories as well. It is a high-end make-do-and-mend approach. Bloomfield suggests that 'sustainability is a mix of ethics, wellbeing and focus. Having time to focus on work you want to do, for customers you want to work with' [3]. He goes on to say that 'this approach contributes to a sustainable working practice. If that work also makes you happy, is enjoyable, improves your skill and understanding then the benefits are obvious. Aligning a beneficial working practice to a sustainable one is then about the choices you make that contribute to an ethical personal perspective' [3]. And Decker has a similar approach but is keen to note that:

being mindful is acknowledging the individual's capacity to engage in a process that is different from everything else, because otherwise we're going to lose that acknowledgement and the recognition for the work that people do. They could be making a wicker basket it doesn't matter. But what matters is to make a small difference in the world [7].

Traditional craftsmen and women operate in a very different world to those employed by luxury brands. They have a command over their work in a way that is not at all possible in the luxury brand environment. There is no denying that an element of skill is required to polish and set stones, but the processes are mechanical, fast and predefined. No creative input is required from the technician who sets a stone or polishes the metal. No intuitive processes are employed as stringent protocols dictate the outcome of the product. There is no slight of hand, no signature of the maker. There is no accountability, no choice. With a craftsman or woman they have the capacity to decide on the materials they use, and they can choose their supplier and build relationships with their customer. As Fennell, Bloomfield and Decker have said, design plays a critical role in 'not only defining the business model but also the products manufacture and materials and technology used' [3]. It is this approach that defines the future of sustainable luxury jewellery. As technological advances impact not only the production of products, it will also contribute to the adoption of transparent practice through the implementation of track and trace, the reduction of mining and the redistribution of reconfigured and reconstituted jewellery. This is the future of a disrupted and more widely distributed supply chain, one that is possibly more ethical, sustainable and desirable.

References

1. Berry C (1994) The idea of luxury. Cambridge University Press, London
2. Berry CJ (1999). Austerity, necessity and luxury. In: Hill J, Lennon C (eds) Luxury and austerity. Dublin, University College Dublin Press
3. Bloomfield M (2020) Luxury jewellery and sustainability (Borstrock S, Interviewer). Accessed 12 May 2020
4. Fennell T (2020) Luxury, craftsmanship and jewellery (Borstrock S, Interviewer). Accessed 18 June 2020
5. Sennett R (2008) The craftsman. Yale University Press, New Haven
6. Pye D (1971) The art and nature of workmanship. The Herbert Press, London
7. Decker U (2020) Luxury and fairtrade jewellery (Borstrock S, Interviewer). Accessed 23 June 2020
8. Fuyu (2020) Mushin, the empty mind. https://www.zenlightenment.net: https://www.zenlightenment.net/mushin/. Accessed 20 June 2020
9. Blume (2020) The history and evolution of the global supply chain. www.blumeglobal.com: https://www.blumeglobal.com/learning/history-of-supply-chain/. Accessed 12 June 2020
10. Pine BJ (1999) Mass customisation - the new frontier in business competition. Boston, Harvard Business School Press
11. Walt (2018) 'Blood gold' is financing rebel armies, drug gangs and child labor. Bloomberg: https://www.bloomberg.com/news/articles/2018-10-25/-blood-gold-drives-an-industry-mine-to-market-transparency-push. Accessed 26 Oct 2018
12. Dhar R (2013) Diamonds are bullshit. www.priceonomics.com: https://priceonomics.com/post/45768546804/diamonds-are-bullshit. Accessed 19 Mar 2013
13. Rapaport (2019) De Beers lowers 2020 production forecast. www.diamonds.net: https://www.diamonds.net/News/NewsItem.aspx?ArticleID=64484&ArticleTitle=De+Beers+Lowers+2020+Production+Forecast. Accessed 11 Dec 2019
14. Warren K (2019) Russia is the world's largest producer of diamonds. I toured a mine in Siberia that produced 1.4 million carats of diamonds in 2018—here's what it was like. www.businessinsider.com: https://www.businessinsider.com/diamond-mine-siberia-russia-alrosa-photos?r=US&IR=T. Accessed 4 Sept 2019
15. De Beers (2020) Provenance and peace of mind. www.debeers.co.uk: https://www.debeers.co.uk/en-gb/provenance-and-peace-of-mind.html. Accessed 14 June 2020
16. Tiffany (2020) Diamond provenance. www.tiffany.co.uk: https://www.tiffany.co.uk/engagement/diamond-provenance/. Accessed 13 June 2020
17. Israel Diamond (2020) Clarity enhanced diamonds. www.israeldiamond.com: https://www.israeldiamond.com/buyer-beware. Accessed 22 June 2020
18. Clean Origin (2020) What is a lab created diamond? www.cleanorigin.com: https://www.cleanorigin.com/about-lab-created-diamonds/. Accessed 24 June 2020
19. Mau D (2019) As lab-grown diamonds near mainstream acceptance, the entire industry is changing. www.fashionista.com: https://fashionista.com/2019/10/lab-grown-diamond-industry-millennials. Accessed 17 Oct 2019
20. E6 (2020) About us. www.e6.com: https://www.e6.com/en/about/corporate-information. Accessed 2 June 2020
21. Responsible Jewellery (2020) Benefits-of-membership. https://www.responsiblejewellery.com: https://www.responsiblejewellery.com/membership/benefits-of-membership/. Accessed 12 June 2020
22. Van Cleef & Arpels (2020) Corporate social responsibility policy. Van Cleef & Arpels: https://www.vancleefarpels.com/ww/en/others/corporate-social-responsibility-policy.html. Accessed 2 June 2020
23. Cartier (2020) Cartier and corporate social responsibility. www.cartier.co.uk: https://www.cartier.co.uk/en-gb/maison/commitments/cartier-and-corporate-social-responsibility/cartier-responsible-jeweller.html. Accessed 23 June 2020

24. Oracle (2020) Improved supply chain visibility and traceability. www.oracle.com: https://www.oracle.com/applications/blockchain/track-and-trace.html. Accessed 22 June 2020
25. Schuster E (2008) Agricultural supply chains: track and trace for improved food safety. http://web.mit.edu: http://web.mit.edu/edmund_w/www/PrecisionAg2008Schuster.pdf
26. Kollewe J (2018) Burberry to stop burning unsold items after green criticism. The Guardian: https://www.theguardian.com/business/2018/sep/06/burberry-to-stop-burning-unsold-items-fur-after-green-criticism. Accessed 6 Sept 2018

Indian Luxury Jewellery—Going #VocalForLocal

Sudeep Chhabra and Ivan Coste-Manière

Abstract Jacques Cartier made his first visit to India in 1911 and developed contact with the many "maharajas" from Kapurthala to Mysore. Most well known of his commissions was the "Patiala Necklace" made for the Maharaja of Patiala with over 2900 diamonds. Inspired by the viewing of the private "Al Thani Collection" of over 250 exquisite pieces from the Indian subcontinent, we questioned if there was a connection between the rich tradition of handmade luxury jewellery and sustainability in India, and is it possible to revive and restore the diversity and richness of handmade jewellery, whilst preserving and sustaining the "artisan". Our research focussed on the launch and commercial success of "Sabyasachi Jewellery" as an independent voice that brought the craftsman centrestage. Our aim was to discover if it is possible to create an ideal balance between commercial success and sustainability of the craftsperson. Our primary method was desk research as well as interviews with customers and independent craftspersons across the jewellery industry in India. We discovered not only the possibility to find this ideal balance between craft and commercial success but a resurgence of customer acceptance and focus towards handmade jewellery.

1 The Mughals and the Maharajas

It was a short walk to "The Grand Palais" from Champs-Élysées, in Paris. The sky was a clear blue with small puffs of white clouds and flowers were in bloom everywhere. It was the first week of April 2017 and we were on our way to "Joy Aux"—an exhibition of jewels from the "Al Thani Collection". This was not ANY ordinary collection. *The Al Thani Collection holds over 6,000 works of art spanning from*

S. Chhabra
Adjunct Faculty, Skema Business School, Valbonne, France
e-mail: sudeep.chhabra-ext@skema.edu

I. Coste-Manière (✉)
Luxury & Fashion Management, Skema Business School, Valbonne, France
e-mail: ivan.costemaniere@skema.edu

I. Coste-Maniere and M. Á. Gardetti (eds.), *Sustainable Luxury and Jewelry*,
Environmental Footprints and Eco-design of Products and Processes,
https://doi.org/10.1007/978-981-16-2454-4_3

the ancient world to the present day. The collection celebrates history and creativity through an exceptional holding of antiquities, jewels, paintings, manuscripts and ancient and historic works of art [1]. We were particularly interested as it was titled "From the Mughals to the Maharajas". The brochure read "*The exhibition presents the major developments in Indian jewellery traditions, from the peak of 17th century Mughal imperial patronage through years of political chaos and colonisation from the 18th century onwards, to the age of the Durbar, great ceremonies that provided Indian sovereigns with a new setting in which to show off their jewels during the time of the British Raj*" [22].

As we exited the venue after an unforgettable three hours, the overarching feeling was of amazement and wonder; of course, mixed with pride and slight anger. We felt anger revisiting the history of how these magnificent pieces of jewels were plundered and passed through generations of Mughals and private jewellers, and also thought of how India was often referred to as the "Sone Ki Chidiya" loosely translated as the "Bird of Gold". But we were also proud of the spirit of creativity and ingenuity of the many local artisans who created these masterpieces. As an Indian involved in the business of fashion and luxury, it seemed relatively easier for Sudeep to understand the evolution of the various techniques of crimping, partial and closed settings of diamonds and the "*polki*" style or the "*kundan*" inlay work that has now become so representative of Indian jewellery. Amin Jaffer, the Senior Curator of the Al Thani Collection, mentioned and quoted: "This collection is unique in that it has a very long chronological period: four centuries. Its diversity is also remarkable: there are both objects that were part of the Mughal Imperial treasure, but also masterpieces of contemporary jewellers who incorporate old stones whose historical value is well known or which draw their inspiration from traditional Indian shapes and patterns" [6].

The fact that we are able to recall each detail of this experience at The Grand Palais even after so many years is testimony to the amazement we had felt when we were there. When we came back to India and wondered who could possibly represent the breadth and width of this amazing history of Indian jewellery, only a handful of names of family-run jewellers came to mind. Unknown to us then, India's acclaimed Indian bridal couture designer Sabyasachi Mukherjee, was also thinking about emeralds, rubies, sapphire and diamonds. Having been obsessed with jewellery since he was a teenager, and after various successful collaborations with the likes of Tanishq, Forevermark [3] and many more, he launched Sabyasachi Jewellery in the fall of the same year.

The first time Sudeep met with Sabyasachi Mukherjee was when Suzy Menkes asked him to address the attendees of the International Herald Tribune Luxury Conference in New Delhi in March 2009 [25]. The theme of the conference was sustainability and we remember Sabyasachi making a passionate appeal to focus on the country's rich craftsmanship and heritage and to go local. Sudeep Chhabra had just taken charge of India's first international luxury shopping centre; DLF Emporio and was eager to know more about the business of fashion and luxury. Whilst it was easy to be enamoured by the big international luxury brands arriving to woo the Indian customer, here was an Indian designer who was fiercely proud of his heritage and refused to adapt to the more western silhouettes that were the hot trend that year. He came across someone as striking clear and almost headstrong, articulating a detailed vision of India and its handcrafts. Perhaps, his was the only address that we recall most vividly. This was the same feeling of pride we had felt at The Grand Palais that April morning in 2017.

2 Sabyasachi and the Launch of Sabyasachi Jewellery

Sabyasachi Mukherjee started his eponymous label in 1999 and tasted success almost overnight. He is India's most successful bridal couture designer, retailing from five flagship stores, employing 1100 people directly and outsourcing work to another 35,000 people. Every Sabyasachi piece is lovingly, painstakingly and deliberately created by skilled artisans using the highest degree of craftsmanship, quality, devotion and care. At Sabyasachi, artistry, creativity and quality are of paramount. Each handcrafted object is unique and special, the making of which involves years of learned skills, passion, enthusiasm, commitment, sincerity, which reflects a living heirloom culture [18]. He calls his design philosophy the "personalised imperfection of the human hand" [30] (Image 1).

Whilst launching Sabyasachi Jewellery, he remarked to his over 4 million fans on Instagram "*Growing up, I spent hours rummaging through my mother and grandmothers' jewellery cupboards. Their taste was simply exquisite. Maybe this is why I feel disappointed. Where has all the art gone? Perhaps it was this question that pushed me to find the answer*" [17]. When interviewed by Architectural Digest,

Image 1 Mukherjee [11, 12]

he said "I am not someone who enjoys prissy jewellery. For me, jewellery can be classical or irreverent or completely obnoxious. All three things for me are exciting—otherwise they don't matter". So, centre stones in charm necklaces are often lower in value than the rarer, and more precious ones flank them; polki diamonds in rani haars (heavy, long necklaces) are hidden under drapes of coloured stones; and the back of every piece is as interesting and exquisitely finished as the front, often carved and engraved with diamonds or lush meenakari (a metal enamelling technique) [19].

The region of South Asia has a very longstanding and unique history of jewellery. *This is partly the result of natural circumstances. The region has been home to gemstones: fine diamonds were found in the Deccan, Kashmir produced sapphires of the most beautiful hue and Badakhshan was home to the most prized spinels. Sapphires and rubies were available from nearby Ceylon (Sri Lanka) and Burma (Myanmar), and pearls were available through trade with the Persian Gulf. Emeralds gravitated to India through a commercial exchange, brought by European merchants after the discovery of mines in Colombia. These precious materials were transformed through the ingenuity of Indian craftsmen, raised to fresh heights by patronage that continues until today* [23] (Images 2 and 3).

One could argue that this geographic advantage is the key reason that India's gems and jewellery sector has been one of the largest in the world, contributing around 29% to the global jewellery consumption [7, 8]. We would propose that it's actually because of the thousands of "karigars" or craftsman as they are called, who have learned the techniques and craft of handmaking jewellery through many generations. India is deemed to be the hub of the global jewellery market because of its low costs and availability of high-skilled labour. It is the world's largest cutting and polishing centre for diamonds, exporting 75% of the world's polished diamonds.

Image 2 Mukherjee [11, 12]

Image 3 Mukherjee [11, 12]

Today, 14 out of the 15 diamonds sold in the world are either polished or cut in India. India exported US$ 18.66 billion worth of cut and polished diamonds in 2019. The sector is home to more than 300,000 gems and jewellery players, employs over 4.64 million people and is expected to employ 8.23 million by 2022 [7, 8] (Images 4 and 5).

Image 4 Mukherjee [10]

Image 5 Mukherjee [10]

3 Jewellery Production: Behind the Doors

Making jewellery has been a vocation-based craft through various generations of goldsmiths, artisans, polishers and jewellers. Till some time ago, it was an unorganised sector concentrated in certain pockets of the country where the whole family was involved in the business. Because of this, certain communities had a majority of artisans and were able to demand a premium for their skills, especially as all the work was done by hand. For example, in a "kundan" literally meaning "24 carat gold" type of setting, the goldsmith sets gemstones and uncut diamonds directly onto embellished or engraved jewels and then uses a very thin foil of pure gold (kundan) that is used to fill the crevices of the embellishment and heat this with small charcoal pieces so as to complete the setting. Not only is this very intricate and fine work, there is a certain skill that comes from practise that cannot be replicated by any machine. Most of these artisans use handmade tools that they have created themselves to make the entire jewellery from start to finish. This often led to a flourishing of creativity so that each family or community evolved their own particular styles and imperfections that contributed to the diversity of designs and craft. Because of this very detailed and slow, painstaking nature of handwork in creating jewellery, oftentimes two pieces were never exactly alike, yet they appeared the same. The amount of time required on each large piece of jewellery was more aligned to commission-based production rather than commercial production.

If you try and look at the huge jewellery market in India, you cannot miss the diversity across different regional cultures. The major consumption of jewellery in Indian households is either for special occasions like marriages and special occasions or for everyday use. A large part of this demand comes from the rising middle-class population whose urban lifestyle influences purchases. Working women prefer comfort as well as flexibility. This has led to a skew towards easy to wear, lighter pieces with more contemporary western influences. As the working middle class moves to urban centres within the country, they often don't live in the same city as their ancestral homes. The industry has thus quickly evolved to mixing regional designs and styles into a blend that is universally more sellable than before. This demand slowly led to the adoption of investment casting and electroforming commercial processes in the manufacture of jewellery, where the need was to have workers who were good at delivering volumes by operating machines. Mechanisation does not require a specialised craftsperson; instead it's preferable to have a worker who is more process-driven. The factory model of investment casting allows many exact replicas of an original design to be mass-produced, rather than hand-made jewellery which would take a long time to make just a couple of pieces. Almost all the major hubs of jewellery production thus moved to casting which was not only more economical as it reduced wastage and was time effective but also reduced defects and labour costs. Whilst companies still employed goldsmiths to oversee parts of design, most of the remaining workers were left to work in process-driven production. Soon, design and manufacturing became two separate entities, as trained designers worked on computers and specialised design software to automatically create fine details with accuracy and speed whilst churning out designs. The quality of machines also improved and the precision with which these designs could be manufactured was even better than handmade jewellery. This led to the slow demise of handmade jewellery. Though many of the signature large pieces were and are still handmade, the majority of each collection was in a large part mechanised. These changing preferences led to an over standardisation in the way the jewellery business is structured. They need to stock a lot of variety and be relevant to a large demographic with very diverse tastes. Beautiful and exotic names of collections, coupled with well-spun marketing messages accosted the average Indian customer who was being exposed to a bewildering choice in the jewellery market. With the advent of e-commerce, and the option to sort on price, mostly the decision to buy was influenced by finding the ideal mix of a number of criteria for each customer. Entry-level discounting, loyalty programmes and exchange offers further fuelled the insatiable demand of the Indian customer who is the largest consumer of gold in the world. India's demand for gold reached 690.4 tonnes in 2019 [7, 8].

As jewellers became more structured in their business operations and started focussing on the customers with large showrooms and a variety of designs, the backend work of designing and manufacturing jewellery started getting aggregated to large manufacturing units. Traditional and "khandani" or "generational" craftsmen slowly started facing unemployment, low income and decreased demand for their skills. This factory model meant that most of the small "karigars" or "artisans" were employed on a contract basis or worked for smaller manufacturing units where specialised work was outsourced to. Because the business is cyclical, this also meant that there was no steady income throughout the year. If you were a polisher or an enameller you were bunched into manufacturing units doing the same work and hence your skill was expendable. Since most of these "karigars" typically started working at an early age within their families, often they did not have time for education. They did not have any formal training or certification that they could rely on to look for work throughout the year. As one grew older, many would have weak eyesight or back problems as their work required them to work in a sitting position for a long time peering down at very fine stones. When work was not available throughout the year in the manufacturing units, many of them started shifting to well-paying factory jobs with large companies completely moving out of the jewellery sector. The following generations aspired to live better lives than their parents, so they never wanted to carry on with the family tradition of being a "karigar". On the contrary, they aspired to have a better education and move to urban areas with better paying and far more stable jobs. This became and continues to be one of the major challenges of the jewellery business today. A large section of craftsmen just disappeared and moved to other industries carrying with them the fine skills and techniques lost to posterity. On the other hand, the jewellery business faced an acute shortage of skilled artisans during high season months. Most of who stayed started asking for a high premium to support themselves for the months they would not have any work. The umbilical cord between generations of artisans was being cut.

These communities of skilled "karigars" were also geographically dispersed in major centres and pockets across the country. Mumbai is an important hub for casting and diamond set jewellery, Thrissur is a hub for southern light-weight jewellery, Hyderabad has a long history of gems, pearls and set jewellery, whilst Jaipur and Bikaner are famous for the "kundan" and "jadau" [26] style of jewellery. Delhi is a large hub for silver jewellery and Surat is world famous for diamonds and polishing stones and Kolkata (former Calcutta) is a major hub for intricate handmade jewellery. Typically, there is a high demand for Bengali goldsmiths for precious metal jewellery, diamond and gemstone polishers from Surat and enamellers from Jaipur, who are mostly paid higher for the quality of craftsmanship they offer. Another unique feature of the Indian jewellery market was that most if not all of the major Indian jewellers actually began as diamond or gem traders (involved in cutting and polishing) and then naturally expanded their business to become jewellery manufacturers. Our personal view is that oftentimes any disruption in a business model comes from outside the business and not from within. So, it is interesting to note that a designer like Sabyasachi who is almost unapologetic in his strong view on the grandeur and opulence of the jewellery business, rooted in tradition and handmade artisanal work, has been able to make a mark on this industry in just a few years. This seems counter-intuitive. When he launched his jewellery line on his Instagram page he said "Indian jewellery is beautiful, arrogant and strong and very individualistic. It's my mission to bring it all back, in all its former glory. And I will not rest till it's done" [16, 26]. Bespoke, at the heart of the vision and the strategy…Very few customers today realise the difference between machine or 3D SLS made jewellery and handmade jewellery and this is why we argue that Sabyasachi Jewellery has made a distinct place for itself. His passion clearly focusses on not what jewellery *is as a product* or accessory but *what it means*. He has been able to stitch the logic of "why" one should buy rather than "what" one should buy. "*Because I say that jewellery should not just*

THE SABYASACHI FLAGSHIP JEWELLERY STORE

Image 6 Mukherjee [11, 12]

be a public luxury, it's a private luxury as well. It is what it does to you, before it does it to anybody else" [19]. In an interview to Vogue he says "The fundamental know-how and the intricate craftsmanship are somewhat lost to the younger generation of urbanised youth. It is the older generation of craftspeople who design with joy. The youth talk statistics and price points. Jewellery needs to go back to its purest form and exuberance" [21] (Image 6).

4 The Kolkata Connection

And herein lays an interesting aspect of geography as well as heritage. Kolkata or Calcutta as it was called earlier has been a major hub for jewellery production, especially known for "filigree" work that requires a lot of skill. Bengal's association with jewellery is a combination of factors, like cost, creativity of design but most importantly immaculate craftsmanship. This has also to do with the continued patronage provided to the craft of the local population over the last many decades. Most of these "karigars" moved to various parts of the country to jewellery manufacturing units as demand for their skills grew and they carried their art with them. And this is what Sabyasachi was able to offer to them but back in their home state. The fact that he looks at the business from the lens of craftsmanship is a critical aspect of his brand ethos. He has been resolute in his efforts to re-create an era of craftsmanship that was long forgotten. As he shares in an interview with Vogue, "I started my jewellery business with the same principle with which I started by clothing business. We have been successful with what we do because we have been able to find a gap in the market that others did not see and bought relevance back to these things", said the designer, whose thought and aesthetic is deep-rooted in old-world Kolkata, or Calcutta, as he prefers calling the city [27, 28]. His focus has been on making small quantities of traditionally designed diamonds and jewellery which will have a higher value and could be worn over many years. Yes, it's more aligned to the heavier occasion-based as well as bridal market that he is known for, but if you look closely; his aesthetic is rooted in tradition. He says, "Lack of patronage and, worse, lack of hope is what was glaring to him on his journey in the world of jewellery. The artisans who have extreme know-how live in abject poverty because they have out-skilled themselves in a market that demands mediocrity. I want to slowly push mediocrity out of the jewellery industry by giving work to the kind of craftspeople you no longer find easily" [21]. When Vogue India interviewed him recently on his twentieth anniversary and asked him about his most important takeaway, he said "The most gratifying thing about being in this business is that I have been able to create opportunities of empowerment for a lot of people working with my brand. Every year I sit with my Human Resources team and recreate a policy where we hire at least 10–15% more people. We hire more craftspeople, we adopt more villagers, and we work with many more craftsmen to be able to create a demand and supply" [27, 28].

5 The Sabyasachi Art Foundation

The Sabyasachi Art Foundation was started by Sabyasachi and his sister Payal, as a tribute to their mother, who used to work in the Government Art College and was deeply involved in handicrafts. It strives to give indigenous artists and craftsmen the due recognition and a means of livelihood. Today, it is an integral part of the brand both in a creative and socially responsible capacity [4, 26]. *To this end, he hires highly skilled artisans from families where the next generation might be veering towards careers in construction or engineering in urban centres, discarding their family businesses, because they have ceased to provide opportunity and value to them. "What this great country offers to you—which we will only realise later, and in retrospect—is craftsmanship", he says. "The way the world is headed, the intangible will become far more important, and far more expensive, than the tangible"* [20]. Of all the people who are in the fashion business, he should know.

The Sabyasachi Art Foundation has 43 artists from West Bengal who need financial support and are unable to market their craft. They have already worked on many an association with Christian Louboutin, a special hand-painted collection of artefacts for Pottery Barn, a long continuing wallpaper collection for Asian Paints and most recently the capsule collection of Sabyasachi with H&M [9]. Sabyasachi has a very deep and respectful relationship within India's artisan community. In fact, he is proud that the entire genesis of his jewellery is based on Bengali craftsmanship. Working with these artisans, he is able to leverage their age-old wisdom and techniques so as to highlight the nuances of traditional designs to build an aesthetic that stands out with a certain character that is his own. He is able to have some of the best artisans who had left to work for the big jewellery houses to come back to Bengal and work with him; because they relate to his ideology of Bengali heritage. He says "*We have started focussing on craftsmanship and craftsmanship is something you have to nurture. We have brought back what used to exist in India 50, 60, 70 years ago, before the great tirade of buying jewellery only for investment happened and completely killed craftsmanship. Thirty years later, if you are not going to save the ecosystem, the second or the third generation might not know how to make it any more, because craft is goes from father to son*" [2].

This unwavering long-term vision that comes from clarity of thought born out of pure conviction is what makes Sabyasachi Jewellery unique. The fact that he designs each piece so that it may be worthy of sitting in a museum in 15–20 years' time is what makes him stand apart. Does jewellery design need to be so individualistic and slightly arrogant? Shouldn't it be more mainstream and appeal to many rather than just a discerning few? This almost counter-intuitive way of running his business is the reason he has carved a place in the jewellery market in the last few years.

As the Made in India movement gathers speed, many designers are focussing on promoting local production to make it sustainable and scalable [13–15]. The Indian government's ***#VocalforLocal*** campaign [5] also encourages enterprises to take a pledge to support local artisans and small businesses to progress together. A large part of the sustainability conversation in fashion is to promote local clusters of craftsmen so that they return to their roots and pass on their expertise and skills to the next generation. One of the ways to encourage this is to not only provide earning capabilities but more importantly a sense of honour and prestige to the craft itself. If any brand is positioned to provide a pedestal to craftsmanship which is synonymous with its design aesthetic, the "truth" of the brand shines out and the customer values this. The new customer of today is beginning to see past the similar sounding marketing campaigns and has started seeking their own reality. As the customer matures in their understanding of fashion and our unique Indian heritage, the conversation has shifted to how the jewellery is made, working conditions and proper wages of staff and what is the brand's stand on issues like social and environmental sustainability. Whilst Indian customers have typically looked at their purchase of gold and diamonds more from an investment point of view and their value appreciation over time, but they seem to be slowly shifting to also what it "means" and the language of design actually helps to uncover that answer. This critical shift is happening slowly, thanks to brands like Sabyasachi Jewellery that espouse craftsmanship and are able to put that on a pedestal. As he says in an interview with Architectural Digest, "The position that we have reached and the power that we have now, the more well-known I am becoming worldwide, the more I am becoming arrogant about my cultural heritage. Bengal had at one time created a cultural renaissance in India, history repeats itself and it's happening again… and while it happens, I want to be the front runner in it" [2]. Sustainable doping on the way… (Images 7 and 8).

Image 7 Mukherjee [13–15]

Image 8 Mukherjee [13–15]

6 The Story of New India

In early 2020, Bergdorf Goodman released the following statement. "As a purveyor of style and taste, Bergdorf Goodman celebrates brands rooted in craftsmanship and innovation. A proud new addition to the department stores featured jewellery brands; Bergdorf welcomed the exclusive launch of Calcutta-based jewellery designer Sabyasachi Mukherjee, who debuted a 65-piece Haute Joaillerie collection featuring all one-of-a-kind pieces" [24]. As the only designer to have a window at the famed 5th Avenue and a curated pop up in early 2020, it certainly cemented this position of Indian handcrafted heritage jewellery. It was very surprising to have a customer be in touch with the store especially during the lockdown to insist to buy a particular piece from the Sabyasachi Jewellery collection for his fiancé. After a virtual viewing was setup by the store, the customer waited for the store to offer kerbside pickup and ended up buying not one but two pieces from the collection. It has been one of our most successful partnerships till date, said Darcy Penick, President, Bergdorf Goodman [29] This first ever virtual sale won't be easily forgotten, especially as it was from an Indian designer who says "success and niceness are not mutually exclusive" [13–15]; we would argue so is the case with jewellery and traditional craftsmanship.

7 Conclusion

When we talk about sustainability, the first thought that usually crosses our mind is that of material and sourcing. In the case of jewellery, designers are consciously thinking of not only this but also the entire process of bringing a piece to life. For those like Sabyasachi, who are able to revive regional craftsmanship and put it on a proud pedestal, it takes an unwavering spirit of conviction to see this process through. And this is also a celebration of sustainability, which is the inclusivity of the local business model. We must highlight this effort which is now certainly finding a valuable place in the minds and heart of the conscious customer. More than an example of a definitive achieved vision, a sustainable strategy....

References

1. Administrator (n.d.) About us: Al Thani Collection. https://thealthanicollection.com/about/. Accessed June 2020
2. Chowdhury SR (2020) Article: The Telegraph Online. https://www.telegraphindia.com/culture/style/sabyasachi-on-his-collab-with-the-iconic-new-york-fashion-house/cid/1740579. Accessed June 2020
3. DiamondWorld (n.d.) News: Diamond World. https://www.diamondworld.net/contentview.aspx?item=15155. Accessed June 2020

4. FDCI (n.d.) Member: Sabyasachi. https://www.fdci.org/members/profile/-1586146406.html. Accessed June 2020
5. Government of India (2020) Pledge: Vocal about Local. https://pledge.mygov.in/vocalaboutlocal/. Accessed June 2020
6. Grand Palais (n.d.) Jewels exhibition: The Grand Palais. https://www.grandpalais.fr/en/node/17505. Accessed June 2020
7. IBEF (n.d.) Gems & jewellery presentation. https://www.ibef.org/industry/gems-and-jewellery-presentation. Accessed June 2020
8. IBEF (n.d.) Indian gems and jewellery report. s.l.: IBEF
9. Khan TN (2020) Article: Entrpreneur.com. https://www.entrepreneur.com/article/345376. Accessed June 2020
10. Mukherjee S (2018) Sabyasachi jewellery. s.l.: Sabyasachi Mukherjee
11. Mukherjee S (2019) Official Instagram account. s.l.: Sabyasachi
12. Mukherjee S (2019) Sabyasachi jewellery. s.l.: Sabyasachi Mukherjee
13. Mukherjee S (2020) Article: Elle India. https://elle.in/article/sabyasachi-mukherjee-linda-fargo-of-bergdorf-goodman/. Accessed June 2020
14. Mukherjee S (2020) Instagram: Sabyasachi. https://www.instagram.com/p/CAHPfRMBMQ-/. Accessed June 2020
15. Mukherjee S (2020) Sabyasachi jewellery. s.l.:Sabyasachi Mukherjee
16. Mukherjee S (n.d.) Instagram: Sabyasachi Mukherjee. https://www.instagram.com/p/BWpqemsFq7h/. Accessed June 2020
17. Sabyasachi Jewellery (n.d.) Sabyasachi jewellery. http://sabyasachijewelry.com/. Accessed June 2020
18. Sabyasachi Official (n.d.) About us: Sabyasachi official Facebook. https://www.facebook.com/pg/sabyaofficial/about/. Accessed June 2020
19. Sanghvi RB (2017) Article: Architectural Digest India. https://www.architecturaldigest.in/content/sabyasachi-mukherjee-mumbai-jewellery-store-design-precious-stones-crystals/#s-cust0. Accessed June 2020
20. Sanghvi RB (2019) Article: Architectural Digest. https://www.architecturaldigest.in/content/sabyasachi-mukherjee-mumbai-jewellery-store-design-precious-stones-crystals/#s-cust0. Accessed June 2020
21. Tewari B (2017) Article: Vogue India. https://www.vogue.in/magazine-story/sabyasachi-mukherjee-fashion-designer-wedding-diamond-jewellery-on-challenges/. Accessed June 2020
22. The Al Thani Collection (n.d.) The Grand Palais Paris: The Al Thani Collection. https://thealthanicollection.com/exhibitions-and-publications/the-grand-palais-paris/. Accessed June 2020
23. The Grand Palais (n.d.) Depliant—Joyaux. https://www.architecturaldigest.in/content/sabyasachi-mukherjee-mumbai-jewellery-store-design-precious-stones-crystals/#s-cust0. Accessed June 2020
24. Town & Country (2020) Article: Town & Country. https://www.townandcountrymag.com/insider/g31004042/bergdorf-goodman-celebrates-the-launch-of-sabyasachi-mukherjee/. Accessed June 2020
25. Two Circles (n.d.) News: Two Circles. http://twocircles.net/2009mar24/go_indian_go_global_fashion_editor_tells_indian_designers.html. Accessed June 2020
26. Utsav Pedia (n.d.) Jadau: Jewellery. https://www.utsavpedia.com/attires/jewelry/jadau/. Accessed June 2020
27. Vaidya R (2017) Article: Vogue India. https://www.vogue.in/content/sabyasachi-mukherjee-just-announced-his-first-ever-jewellery-collection. Accessed June 2020
28. Vaidya R (2017) Article: Vogue India. https://www.vogue.in/content/designer-sabyasachi-mukherjee-20-year-anniversary-exclusive-interview. Accessed June 2020
29. Vaidya R (2020) Article: Vogue India. https://www.vogue.in/fashion/content/sabyasachi-and-bergdorf-goodman-make-a-virtual-sale-of-stunning-jewellery-defining-kerbside-luxury. Accessed June 2020
30. Wikipedia (n.d.) Wikipedia: Sabyasachi Mukherjee. https://en.wikipedia.org/wiki/Sabyasachi_Mukherjee. Accessed June 2020

Viable Pearls and Seashells: Marine Culture and Sustainable Luxury in Broome, Western Australia

Annette Condello

Abstract This chapter traces the origin of pearls and seashells as underwater luxury goods from a global-historical and marine-cultural perspective within the sustainability debate. It illuminates the luxury of the hidden sea gems, its natural and imitation items, refers to writings concerning jewellery ethnography and discusses the impact of the connections between the pearl, fashion and tourism industries in Western Australia. Contributing to creating Broome's viable marine-culture, it analyses the popularity of the Australian pearl and pearl shell overseas as well as relooking at the jewellery traders who introduced the luxury of the indigenous "ritual good" to an international market in the context of "sustainable luxury". The chapter concentrates on the display of the Western Australian indigenous commodity in newspapers and magazines from the 1930s to the 1960s. Suggesting how the indigenous integration of seashell luxury emerged in Australia's Northwest, the town of Broome impacted the multicultural pearling industry not only through jewellery, but also by providing an alternative solution for pearl fishery farms. The viability of the pearl industry demonstrates how the natural jewel has contributed to the meaning of sustainable luxury, transforming how to measure coastal remnants in the Kimberley region. Broome's marine culture developed a form of Australian pearlescence as a sustainable luxury, with Paspaley and Kailis jewellery injected with authenticity, reliability and organic substance.

In the era of excessive consumerism unlike any other piece of jewellery, the pearl is an organic sea gem created from a single-living mollusc on the sea floor. With growing interest in artisanal entrepreneurship, pearls continue to be a popular gem choice for both men and women. Since antiquity, pearls were recorded by Roman general, Pompey, as one of the richest luxury commodities to own [24], as is the case today with naturally sourced ones. Owing to their large size, south sea gems

A. Condello (✉)
School of Design and the Built Environment, Faculty of Humanities, Curtin University, U1987 Perth, WA, Australia
e-mail: A.Condello@curtin.edu.au

I. Coste-Maniere and M. Á. Gardetti (eds.), *Sustainable Luxury and Jewelry*, Environmental Footprints and Eco-design of Products and Processes, https://doi.org/10.1007/978-981-16-2454-4_4

that increase their lustre through the intensity of light rays and the surplus of their oyster shells have become viable jewellery types, revered and rated highly globally, particularly in Western Australia.

Renowned globally as a pearling and trading port in the early 1900s, Broome in the Kimberley region in Western Australia became the multicultural pearling capital of the world. In terms of ancient indigenous cultural-exchange in this particular region with the Indonesian archipelago, however, diving for seashells predated more than a few thousand years before. In the coastal cosmopolitan town of Broome in particular, what followed was the incline of the pearl industry in the early 1910s. This incline also prompted an invention and integration of indigenous luxury in Australian communities that involved engraving seashells for bodily ornaments. Broome's marine culture developed into a sustainable luxury venture. It has attracted more consumers to purchase south sea pearl jewellery.

This chapter traces the origin of the hidden sea pearls and seashells as underwater luxury goods from a global-historical and marine-cultural perspective within the sustainability debate. In tracing what constitutes "sustainable luxury", considering both natural and imitation items, the chapter refers to writings concerning jewellery ethnography and more broadly the impact of the connections between the pearl, fashion and tourism industries in Western Australia. These connections, in turn, contribute to creating Broome's viable marine culture. Analysing the popularity of the Australian pearl and pearl shell overseas and the jewellery traders who introduced the luxury of the indigenous "ritual good" to an international market, the chapter observes the display of the Western Australian indigenous commodity in newspapers and magazines from the 1930s to the 1960s.

1 Pearls and Seashells as Underwater Luxury Goods

> "Luxury was perhaps born on a day of foaming seas, when spume trims waves with a lace of water that dances as they ebb and flow. Since the dawn of time, like the oceans, in an eternal renewal, it has become, in turn, rare, precious, innovative, unique, multiple, extravagant, minimal" [15, p. 198].

We wear pearls to ornament our body or to fasten clothing, and the oysters are a source of deluxe food. Renowned mythically for leading excessive lifestyles, the ancient Sybarites revelled in dining on oysters and wore pearl necklaces at their banquets [8]. For Roman Emperor Heliogabalus, for instance, in lieu of spice he relished in his sybaritic habit of sprinkling pearl grindings on oysters for enjoyment as well as for nutrition. Spent shells, too, are valued for their nacre, the physical inner covering of pearls otherwise recognised as mother-of-pearl (or madre perla). "Nacre is formed by the mantle of the oyster-shell, which extracts lime or calcium carbonate from the water" [34, p. 23]. The oyster "protects the soft flesh of the oyster from the rough shell. When a foreign object or parasite such as a bit of broken shell or grain of sand or a tiny worm, crab or fish gets inside the oyster-shell, the mantle secretes the nacre around it again to protect the soft flesh" [34, p. 23]. Sourced by diving to the bottom of the ocean floor, shucked and cleaned, and created into jewellery, pearls and their shells are valuable because they are organic. They also enabled the manufacture of mother-or-pearl buttons and glass beads in Middle Eastern, Asian and European cultures. Deriving from the nacre from the shells, lustre creates underwater luxury.

Fashioned from nature, the naturally cultivated seashells and their pearls offer an abundance of underwater luxury goods in the sense their excessive matter is ornamental and yet sustainable. Once reserved for the wealthy, practically anyone now is able to ornament or consume pearls in one form or another whether they are real or fake. Moreover, for centuries artisans have imitated pearl oysters for jewellery and seafood for nutrition. In antiquity, a mix of "mica, wax, eggwhite, and mercury held together by gum tragacanth" (Quoted Schorle [30], footnote 59, 53) created fake pearls. Whereas in Renaissance Europe, fake pearls were made from an array of glass beads, which were lacquered with ground fish scales to create a simulated iridescence.

On the one hand, from knife- to gun-handles, to caviar spoons, layers of mother-of-pearl rather than its mass of memories have provided strange ornate coverings for objects for ritual purposes, intruding or improving the environment for people's survival. A late 1980s report in the *New York Times Newspaper* mentions that once these highly valued sea products came from North Western Australia: "every button that graced every starched shirt at every dinner table between St Petersburg and San Francisco came from Broome. Knife handles, hairbrushes, barrettes, cigarette cases—if they were plated with mother-of-pearl they almost certainly had their provenance in Broome" [41]. On the other hand, in the 1930s some pearl products were considered as cheap goods, especially during a sea-jewel glut. In every can of oysters, for example, a genuine pearl could be found, created for the American market.

1.1 Illuminating the Luxury of the Hidden Sea Gems

In illuminating the history of the hidden sea gems, one of the most alluring objects on display at the *10,000 years of Luxury* exhibition held at the Louvre Abu Dhabi, United Arab Emirates, in 2020 was a solitary Abu Dhabi pearl. As a precious organic commodity, in the Neolithic period at Marawah Island in the Emirate of Abu Dhabi, excavations conducted in 2012 have uncovered burial chambers where a natural pearl dating around 5,800 BCE was discovered (see Fig. 1), which was believed to be "the world's oldest known pearl" [15, p. 18]. In 1300, Inb Battuta, for example, recounted his experiences of observing an abundance of such sea gems at pearl fisheries in the Persian Gulf [20].

Offering an alternative interpretation on how ancient luxury is observed in the Middle East, what is curious is that in the Arabic language there is no equivalent term for the Latin term "luxury", only in the sense of it relating to "refinement, ostentation or wastefulness" [15, pp. 16–17]. In terms of characteristics of these living things, the fourteenth-century philosopher Ibn Khaldun "considered that 'the

Fig. 1 Abu Dhabi pearl found at Marawah Island, United Arab Emirates, ca 5800 BCE exhibited at the *10,000 Years of Luxury*, Louvre Abu Dhabi. *Photo* A. Condello 2020

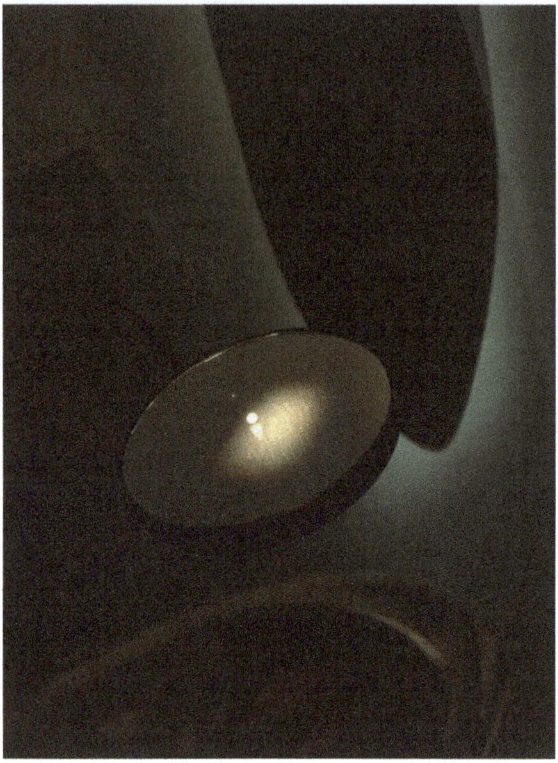

ultimate aim of any civilization is luxury" (Noujaim cites The Muqaddimah, in [15, p. 16]). "Once this goal had been achieved", however, "civilization becomes spoilt and declines, just like all living things" (Noujaim cites The Muqaddimah, in [15, p.16]) but not the strong-worthy Abu Dhabi pearl.

More importantly, "To speak of luxury in art history is therefore a subtle distinction, which probably did not exist in the distant past. Each civilization has its own codes, canons and identifiers. From Ur to Palmyra, from Giza to Rome, in the Middle Kingdom as in the Inca cities, in Polynesia or among the Aborigines of Australia, different techniques, decorations and materials…luminous shells…offer so many different approaches to the fragility of luxury, of what is rare, precious and singular" (Souraya Noujaim in [15, pp. 16–17]). Rather than the pearl per se, the concern for seashells in indigenous communities formulated their own ritualistic luxury codes, mainly in Australia and the South Pacific as a whole. According to Dreamtime, stories and traditions, through the engraving of the luminous shells and in terms of their ritual power, in northern Western Australia Indigenous Elders believe the lustre shells are worth more than the pearl oyster gems. For Yawuru Leader Patrick Dodson, "it must not be forgotten that from time immemorial, our Peoples have harvested pearlshell and traded it across the continent as ritual goods of high value" (quoted from Edwards and Yu [12, p. 6]).

The excessive characteristics of pearl oysters as underwater luxury objects would lend themselves more towards natural ostentation. "Pearl oysters come from a remote branch of the oyster family called *Pteriidae*" [34, p. 23]. Today, according to Pierre-Alexis Dumas, "at a time when luxury seems to hold omnipresent status, in the media and in social networks, it is perhaps pertinent to recall that the luxury object is precisely one that is preserved and passed on—a magnificent, enduring conduit of culture and knowledge, bearing witness to human ingenuity" [15, p. 15]. Indicating the pearl as a preserved and passed-on ostentatious jewel, embedded with culture as opposed to a cultured pearl variety, its long life as a sea gem and as an underwater luxury good is above all precious and possesses combined characteristics that make up a pearl oysters.

Possessed with the strong lustre of underwater luxury and associated moral attributes, authentic natural pearls and the inner lining of seashells are associated with "refractory devices" [28]. In response to the consumerism debate and as a reflection of personal goods to show the world, pearls contribute to a different form of displaced luxury, meaning they are derived from another exotic locale [8] in the sense they are valuable underwater things, sea gems. "A thing", however, "can be a luxury only if its possessor experiences it, through an operationally-oriented expression of his/her own desire to possess, as something that exceeds, in effort, both what is technically necessary for something and what is anthropologically necessary for someone" [40, p. 107]. Apart from its anthropological necessity for somebody, the pearl, for eighteenth-century French philosopher and moralist Jean de la Bruyere, is associated with one of the world's "rarest things" close to "sound judgement" [10]. As rare underwater luxury "things" are enriched with moral attributes, pearls as small goods for personal use also drive contemporary fashion jewellery by exemplifying an amalgamation of nature and marine culture.

In antiquity, pearls gave rise to a range of myths that were associated with the marine-cultural environment. Concerning the specific characteristics of natural jewellery and marine culture, for historian Marcia Pointon, up until the modern era, the origin of precious pearls from the ocean to form a string around the neck (see Fig. 2) gave rise to a range of myths in the ancient Middle East and Asia. Ancient Middle Eastern people "believed pearls were formed from droplets of moisture, the tears of the gods, falling into open oysters which then snapped shut" [28, p. 114]. Within the Indian sculpture, "the gods are associated with pearls. In China and Japan, pearls were thought to result from a combination of thunder and moonlight" [28, p. 114]. "Pearl formation myths" for both cultures "attribute this remarkably beautiful gem to a combination of light and water (the moon and the sea)" [28, p. 114].

Ranges of ancient Middle Eastern and Indian pearl myths connect with the rise of the market demand for pearls for jewellery, especially in the Roman Empire. This was the case with the depiction of pearls within Roman building-shells as well. At a time when Roman elites and merchants were involved in the Indian Ocean trade, what archaeologists found within the House of Meander in Pompeii was evidence, evoking how a bundle of pearls existed at an excavated house in the second half of

Fig. 2 Antique necklace from India ca1880. Natural gulf pearls, mutlicoloured enamel and white stones, exhibited at the *10,000 Years of Luxury*, Louvre Abu Dhabi. *Photo* A. Condello 2020

the first century BCE. "Mosaics in Insula v.i.15.14 in Pompeii, for example, depict a matron wearing a gold and pearl necklace and pearl earrings" (see Fig. 3).

Fig. 3 Ancient roman fresco at Pompeii's house of venus in the shell wearing jewellery. *Photo* https://commons.wikimedia.org/wiki/File:Pompeji_Venus_FoNo.jpg

Luxury and Middle Eastern jewellery, too, "became intricately woven into the fabric of the early Imperial literature and poetic writings of the Golden Age, particularly in love elegies" [30, p. 44]. Other evidence shows a Roman fresco at Pompeii's peristyle of the House of Venus in the Shell, revealing a goddess emerging from the ocean from the shell at birth, wearing jewellery around her neck, ankles and wrists. Such an image no doubt would have been derived from the luxurious lifestyles of the ancient Sybarites, as indicated in Athenaeus' writings [8]. More specifically art-historical writings suggest that pearls were evident in the Roman Empire mainly via the eastern Mediterranean, from Egypt, Syria and the Persian Gulf. "Sculptural portraits of Palmyrene women show them wearing jewellery…. The light weight of pearls, as with silk, made them particularly suitable for long distance trade across difficult terrains, and a likely source for good profits for merchants" [30, p. 44]. Portable pearls are thus talismanic and they embody "apotropaic qualities" [28].

From the association of authentic pearls as morally sound things to symbolically being charged by the gods of the atmosphere to their presence in love elegies, the depiction of them within architectural chambers props them as significant underwater luxury goods tainted, though, with the organic and natural changes through time. Such underwater luxury goods provoked ancient Roman scholar Pliny in his Natural History, Book IX. Pliny would otherwise argue that the representation of pearls was an attack against luxury. According to Pointon:

> Pliny discusses pearls in the context of a diatribe against luxury. An examination of the biological processes through which a pearl is produced enables Pliny to draw a series of connections between the human body (*totum corpus anima hominis*) and the desire for luxury, which endangers not merely the individual but most importantly the nation. The generative powers of oysters and the capacity of pearls to mirror natural changes in the universe….
>
> If it thunders, Pliny says, 'wind pearls' are produced. These are 'only inflated with an empty, unsubstantial show: these are the pearls 'miscarriages. If a shell sees a hand, it shuts up and conceals its treasures [28, p. 117].

Apart from the natural opening and closure of a shell with its pearl(s), the entire world in front of someone is captured and represented in each sea gem into infinitesimal global mirrors. They also mirror natural changes that contribute to the jewel's appeal as a reliable organic gem as the one that is not faceted but always remains the same and only changes its surface when light changes its curvature.

In addition, Pliny also recorded that "pearls in general as the most expensive products of the sea, and he labels gems 'the supreme and perfect aesthetic experience of the wonders of Nature.' Thus, when displayed in the triumphs, the gems and pearls were not only spoils of extreme wealth and royal origin. They also connoted Roman mastery over Nature itself" [24]. Nature, therefore, is what makes pearls so alluring and yet beguiling because of their universal luxury pearlescence, alluding to the origin of sustainable luxury from the Middle Eastern sea. Whereas, Aboriginal art would otherwise indicate the pearl shell and their mastery of it as the original identifier of sustainable luxury because of the ritualistic purposes of the pear shell ornaments on their bodies.

2 Sustainable Luxury, the Jewellery Sector and Shell Mania

Currently, there is a rise in research concerning jewellery as a luxury and its association with sustainability. Reflecting upon writings about the circulation of pearls, in particular the seashells on the international market from the Renaissance era onward, when seashells became excessively ornamental they enable us to understand how they have become more of a sustainable venture as personal possessions. This is especially the case when observing the shells and cement covering Baroque or Rococo jewellery boxes and chests. Today in particular, the jewellery sector promotes a "sustainable luxury practice". For Cappellieri, Tenuta and Testa:

> Luxury products, including jewellery, are sustainable for three main reasons: they represent a crucial sector for the global economy...; luxury helps to put into practice one of the principles of the ethical consumer, i.e. to buy goods by focusing not on quantity but on quality, thus buying in a more responsible manner...; true luxury is sustainable and respects people in the environment, taking also into consideration that a brand's reputation increasingly passes through sustainability [5, p. 6].

However, there is a lack of research explaining the significance of pearl shell ornament as equally as an important natural resource rather than second-rate sea products. Nonetheless, taking on board this responsible manner put toward in the above statement puts the jewels literally on the line, with its diverse qualities of natural or cultured pearls and seashells, which present unique cases of sustainable luxury as in the case at Broome with its reputable global stance as the best the Indian Ocean can offer. Universally, the craze for such sustainable pieces of jewellery and mother-of-pearl buttons became quite popular because of its abundance from the sea.

Jewellery ethnographers have suggested the sea gem is a "mnemonic device rooted in fantasy" [33, p. 66]. For them, jewellery are "treasures that are displayed or hidden in chests, cases and boxes, jewellery activates a memory which is primarily subconscious, and speaks both of presence and absence, of secrecy and disclosure, of the forgotten and the 'souvenir'" [33, p. 66]. As memorable and keepsake luxury commodities, pearls were, and still are, "low bulk, high value, and relatively easy to acquire and transport. Great profits could be made from relatively small quantities of pearls, which have been the added advantage of not spoiling. Furthermore, pearl fishing did not require a large outlay of capital" [17, p. 46]. To be valuable, pearls do not require any type of manufacturing [17], that is, if they are not fake or take the form of precious stones mined and prized from rock, like pink diamonds or black opals.

In Renaissance Europe, pearls became eminent sea gems to own. "In some Renaissance treatises on precious stones, the gem that displaced all stones during the period was the pearl" [9, p. 163]. Pearls, however, "went in and out of fashion in Europe and India with striking regularity" [17, p. 46]. However, when pearls became unfashionable, the sea gem market continued to flourish elsewhere, mainly the Central and South America. Signs of this flourishing pearl market are apparent when viewing Baroque portraits of European nobles or the Mughal rulers in India wearing strings of pearls or garments sprinkled with stitched pearls. At that time "the vast majority of the world's pearls historically came from the Southern Persian and Arabian Gulf" [17, p. 46].

Evidence suggests that sixteenth-century Venetian explorer and jeweller Gaspari Balbi who travelled to the Middle East mentioned the islands that now exist in the Abu Dhabi Emirate "were a main hub for pearling" [15]. Arguably, Balbi's exploration of Abu Dhabi's islands influenced Venice's creative jeweller's skills, as well as those skills deriving from Africa, influencing the origin of Europe's glass beads and button manufacturing.

Through the process of manufacturing glass beads with "the essence of the Orient", at the same time the imitation of pearls started a roaring trade, especially in Italy with the establishment of the "Perleri" or glass bead artisans. In seventeenth-century France, glass beads were inserted with fish slime since there had been a demand for the production of them with pearlescent coatings. During the eighteenth century, the well-established Venetian glass button and innovative pearl manufacture "was mainly entrusted to *perleri,* who represented one branch of the ancient Venetian glass guild.... The vast range of accessories produced by these artisans—including fake pearls, was made using the heat of a lamp to melt coloured-class canes" [2, P. 700]. "*Perleri* were able to improve their collection of samples by combining the use of coloured glass canes with lumps of vitreous paste" [2, p. 701]. Following the *Perleri* trade of the manufacturing of fake pearls, came along shell-mania as well. Shell cabinets of curiosity incorporating mother-of-pearl were popular throughout Europe as with the Rococo period. Conchology, the craze for collecting seashells encrusting jewellery boxes and cabinets of curiosity, became a naturally creative pastime for people in nineteenth-century Britain and linked to sustainable luxury and that extended to the extraction of crude oil and natural jewellery.

Meanwhile, in late nineteenth-century northern Australia, when the shell market raised, Marcus Samuel, for example, sold shell trinkets in the United Kingdom. Considered by some as a "shell merchant" who eventually established the "Shell Transport and Trading Company," his son, while looking for oil in the Caspian Sea, used the mussel shell as a logo, subsequently replaced by a scallop shell [39]. Curiously, the scallop shell advertisement promoted a collector to signpost Western Australia wide variety of seashells and prompted tourism in the northwest pearling regions in the early 1960s, suggesting contradictory notions of that particular form of sustainable luxury in advertising, mining and exploitation by depleting some of the regions' natural resources (Fig. 4).

The need to reuse old pearls and mother-of-pearl buttons in the twentieth century implies an early form of sustainable practice as well as creating or culturing new types of jewellery for the global market. On the fashion and cinematic screen scenes, for instance, in the 1920s Rudolph Valentino dressed in pearls for the Young Rajah film; Erte's clothing designs were covered with carefully stitched-on pearl buttons; and African-American dancer Josephine Baker "wore strings of cultured pearls... on the Paris stage" [34]. In terms of imitating real ones with the invention of Bakelite in the Plastic Age, the manufacturing of "Galalith pearls", which were a reference to Greek pearls known as "milk stones" because they were associated with proteins

Fig. 4 Shell touring service advertisement showing different types of Western Australian shells. *Credit The Australian women's weekly*, 29 March 1961, p.58 / Are Media Pty Limited / aremediasyndication.com.au / and National Library of Australia [NLAref138936]

found in dairy products, became cheap to produce since were innovative in the fashion industry [18]. No doubt, the Galalith pearl was derived from Philostratus' second-century AD writing when he wrote about "how divers would pierce pearls with a long pin to extract a white liquid which they collected in small iron moulds [12, p. 105]. Galalith pearls made out of Bakelite were, produced in Australia in the 1930s, influencing Sydney's fashion industry (Fig. 5).

These industrious feats also prompted Japanese pearl divers to begin the practice of culturing pearls as well as the manufacture of artificial pearls from plastic. In addition, after the post-war bombing of Pearl Harbour was the craze for plastic pearl-strings on

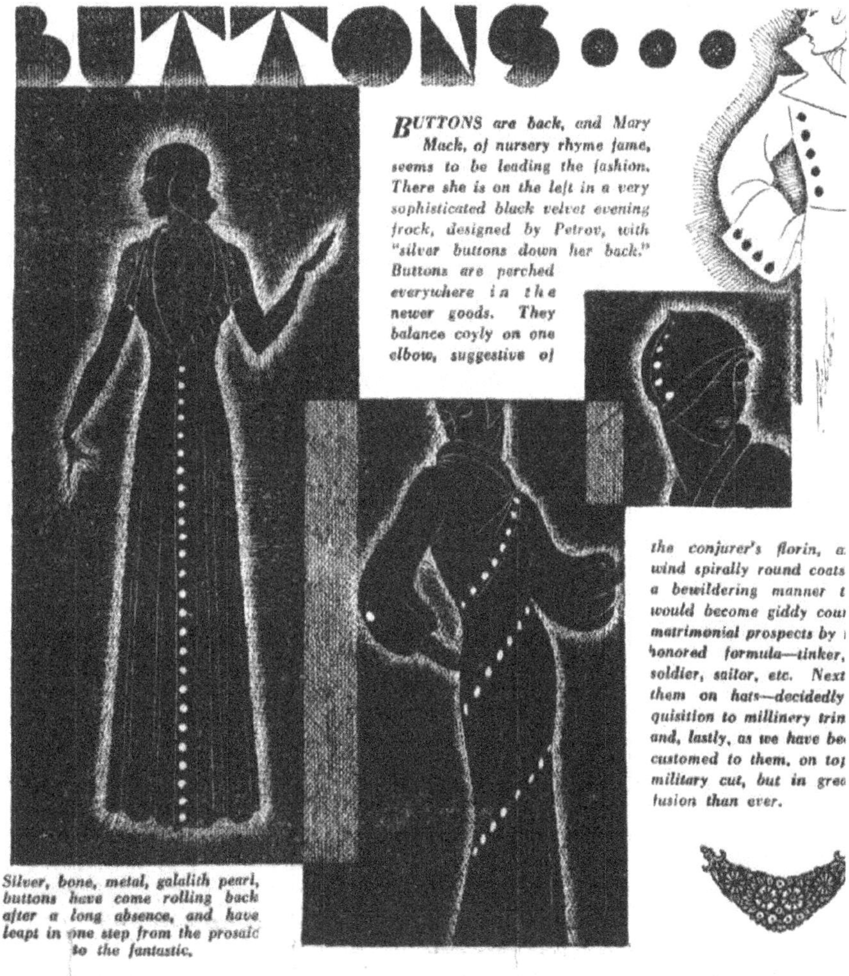

Fig. 5 Galalith pearl buttons. *Credit Smith's weekly*, 20 August 1932, p.20 / National Library of Australia [NLAref138936]

the luxury market. When visiting websites selling quality vintage goods, one might encounter both reused old pearls and mother-of-pearl buttons (including fakes ones), which has most definitely become a sustainable luxury practice, changing strings neck-and-neck.

3 Indigenous Integration of Seashell Luxury in Australia's Northwest

Turning to Australia, the ancient Indigenous peoples from the Northwest traded with peoples from the Indonesian archipelagos well before the arrival of Europeans since they first engraved oyster shell. "For at least 30,000 years, Aboriginal people have carried pearl shell and cone shell beads extraordinary distances from the coast" [12, p. 34]. Excavated in a Kimberley rock shelter in northern Western Australia, a small fragment of the nacreous shell is believed by some archaeologists to be the "oldest Australian evidence that has traditionally been greatly valued by Aboriginal people" [12, p. 34], that is, beyond the age of the found Abu Dhabi pearl.

Interestingly, evidence suggests that "oyster shell meat was a traditional staple for many of the coastal Aboriginal groups and the shell was used and trade with other groups" [19]. As far as indigenous integration of sustainable luxury [7] is concerned, the organic material "was carved, decorated and worn by Aboriginal warriors to show that they had reached manhood" [19, p. 53]. What is ironic is that "Aboriginal people did not desire or value pearls; their only interest was in the shell and shell meat" [19, p. 53], that was, and still is, a vital delicacy for some.

Kwaymullina [19] then claims that the indigenous peoples of Australia were the world's first pearlers. Regardless of which culture was first, it appears the Indigenous practice of placing more significance on the shell is rather curious indeed: "pearl shell is a vital element in Central Desert rainmaking ceremonies" (Edwards and Yu [12], 32).

There was also a concern for pearl shell in the indigenous ritual involvement. Shell gatherers would find the organic material and use the lustre shell part for crafting the "Riji" (or carved pearl shell) to change the weather conditions to suit their needs (Fig. 6). For Moore,

> biological reproduction was not threatened. Consumption of the shell was also influenced by mythology, tradition, ceremonial and religious practices. Pearl shell was given a special status by its selective use in initiation ceremonies to cure illness and to arrive at justice and was usually used by specialists. Even today it continues to form an important part of the socio-economic exchange systems which continue today to operate in Northern Western Australia and the Northern Territory. Overall the Aboriginal industry operated at a low level of exploitation through its collection and consumption practices.
>
> By comparison, the European industry grew quickly to satisfy demand for buttons and fancy items in the Victorian style. At various times it exhausted known shell supplies [22, p. 124].

Additionally, the indigenous pearling industry was considered as a small operation, however, it "involved exchange with other Aboriginals inland and South Eastern

Fig. 6 Riji or pearl shell ornament with ochred engravings, exhibited at Australia Musuem. *Photo* https://upload.wikimedia.org/wikipedia/commons/6/60/Riji.JPG

Asian trade. In common with other indigenous fisheries, the level of exploitation was relatively low and was influenced by animistic ritual. Aboriginals harvested shell by walking on exposed reefs during low equinoxial tides. Shell in deeper waters was not harvested" [22, p. 124].

Moreover, along the endless beaches at Broome "pearl shell" was used in crafting indigenous jewellery. The actual town of Broome "is named after the telegraph cable that ran between Australia and Java, and came ashore there. Cable Beach is vast…and the town local people call it the Port of Pearls" [41]. Concerning pearls, "Australia has an international reputation for the best and largest pearls in the world. The success of the pearling industry is attributed to Europeans who are said to have survived the harsh north-western conditions to settle the Kimberley and the Pilbara, and to establish the lucrative pearling industry we see today" [19, p. 53]. Within the pearling industry, there were other problems. This natural jewellery industry "was built on the exploitation and inhuman treatment of Aboriginal people" and they were originally the ones "who harvested the 'queen of gems' and insured Western Australia's reputation as source of the world's finest pearls. Yet they receive no credit and usually no mention" [19, p. 53], but the community residing in Saltwater Country has recently been credited. This was the case when one of Australia's oldest industries evolved through seashell engravings exhibited at the 2015 *Lustre: Pearling and Australia* show at the Western Australian Maritime Museum.

4 The Multicultural Pearling Industry's Impact on Broome

Recognized as a pearling and trading port in the early 1900s, the town of Broome in the Kimberley region, Western Australia, became renowned pearling capital of the world. Broome "has always been known as a strange, vaguely surreal place, a sort of eastern edition of Cannery Row" [41], that is, similar to a waterfront street in Monterey, California. "By 1900 Broome's docks were thronged with Malays and Chinese and Indonesians, with Polynesians and Japanese" [41, p. 15]. According to Paterson and Veth: "pearl fisheries were crucial to the history of northern Australia, forming a three-thousand-mile chain of discontinuous fisheries from Shark Bay in Western Australia to Cooktown on the Pacific Ocean. The pearl fisheries were Western Australia's most valuable colonial export" [26, p. 3]. They continue: "the earliest reported pearling fields were at Shark Bay in the 1850s, followed by Cossack and nearby Roebourne… and from the late 1860s pearlers shifted from collecting coastal shell to diving, a dangerous job Aboriginal men and women excelled at" [26, pp. 3–4]. It was not, however, without other dangerous or economic problems.

In regard to the economic management of pearl shell in Western Australia from the 1860s to the 1930s, Moore claims that the,

> "demand for shell and pearls became a major problem from the onset of the First World War until the end of the 1930s. The War brought about an immediate contraction of the markets, which did not fully recover post-war as the world economy slid into recession. Process for both pearls and pearl shell collapsed further as the global economy slid into the 1930s depression and technological changes produced cheaper synthetic substitutes, which led to a shift in a demand away from natural products. Plastics came to take the place of pearl shell in buttons and cheaper Japanese cultured pearls replaced Australian natural pearls" [22, p. 121]

Since it takes at least two years to make a flawless cultured pearl.

In the 1930s, pearl culturing was prohibited in Western Australia because of the outcome of over-production. Curiously, in 1937, according to a prominent Broome pearler AC Gregory, there was a novel marketing scheme invented: an experimental way of disposing of cultured cheap pearls through selling canned oysters. One reporter recorded they were "taken from the pearl fisheries around the coast of Japan" and each can of oysters comprised a genuine cultured pearl. They were "examined by a secret process to ensure that a pearl is inside the shell and then packed within 12 hours" [16, p. 2]. This was one way to relieve the glut of these sea jewels. The majority of these canned fishy jewels were exported to New York. This marketing scheme offered the world of gastronomy and luxury consumers "an 'easy instalment' way of collecting pearls for a necklace or other ornament'". More than that, this form of modern marine culture offered a viable way of promoting cultured pearls and food source as doubly informed sustainable luxury. With the impact of Japanese cultured pearls and "the discovery of oil in the Middle East, the pearl industry began to fade. It was considered too dangerous, too strenuous, and the prices that pearls were commanding on the world markets declined precipitously as the Great Depression focused people's attention on daily necessities rather than luxury items" [17, 49].

By the late 1930s, a big discovery was found in the Indian Ocean, one of the largest finds in Australia, if not the world. "Worth 5,000 times its weight in gold, has brought new life to the industry… The biggest pearl ever found at Broome, a pigeon's egg of roseate colour, weighed 160 grains. It brought only $20,000, for the demand was poor at the time…. Seed pearl, flat button, high button, double button, pear or drop and perfect round—this ascending scale of pearl excellence" [1, p. 18].

In 1950, Broome was at the forefront of diving for natural pearls and producing the finest sea gems. Not only the pearlers, or divers, and traders brought pearls to the consumer market from the environment, but also the United States whaling industry in the north. Fashion and tourism followed later. In 1959, New York models flew to Broome for a "pearly" fashion shoot and used the natural landscape by the Indian Ocean as the backdrop, similar to how Paspaley jewellers use the natural topography to sell sustainable luxury jewellery (Figs. 7 and 8). And a year later, Shell Oil Company advertised travel to Australia's northwest and featured a page of unique shells of the state in *The Australian Women's Weekly* (1961), promoting Broome to the world.

Notwithstanding these conditions affecting Broome's pearling industry, there is a greater emphasis today placed upon the significance of the indigenous connection in the Kimberley region to pearl jewellery and sustainable luxury. In addition, Broome's demand for such precious and viable natural products is growing.

Apart from this growing demand, in terms of how these jewels are sourced, there are strategies that have further developed the industry into a more sustainable practice through the Kimberley process but only for pink diamonds. Although "within the jewellery arena, ethical sourcing of gemstones and precious metals has gained some traction, as demonstrated by the Kimberley Process, an international initiative to stem the flow of conflict diamonds (Haufler 2010)" [23, p. 78], these authors specifically refer this process to diamonds. It excludes pearls. Similarly, "the precious stones sector has also taken an active role in promoting sustainability…by the Kimberley Process. The certification agreement developed and approved with joint effort of the government of many countries, of multinationals diamond producers, and civil society" [5, p. 9]. Sustainable luxury and pearl jewellery surviving in other locations (specifically beneath the waters of Palawan Island in the Philippines) presents an ongoing debate.

Real pearls characterised the northwest Australian pearl industry. Paspaley and Kailis Jewellers grade pearls in terms of their size, shape and colour and safeguard the pearl-bearing oysters of Broome, amongst other Western Australia pearlers (Fig. 9). While the white variety is more valuable because English royalty preferred them this way, in Asia the reverse is the case where the golden variety is much more prized. Pearls remaining in the water for longer periods influence the pearl rating. In the mid-2000s, however, a mysterious disease struck the Australian pearl industry and there was a reduced demand for them. To date, the future of pearls requires the producers to rear disease-resistant pearl oysters. Nonetheless, over the years there has been a demand for coloured pearls, which have increased and attracted a different appeal for the luxury jewel market.

'By the dawn's early light . . .'

Americans' 4 a.m. stint to get some pearly fashion pictures at Broome

ROCKS near Broome provide a background for "Seventeen" models. Photographers worked in the dawn light to give these pictures the "pearly" look.

IN their search for unusual backgrounds for fashion pictures, a team of two top New York models, three photographers, and two fashion experts from America's teenage magazine "Seventeen" travelled 12,500 miles to Western Australia.

They took the pictures on these pages at Broome, pearling industry centre, on the remote north-west coast.

The theme of the series is "the pearly look," with extensive use of buttons, jewellery, and ornaments made of Australian pearl shell. These are teamed with summer fashions.

The picture on our cover was taken in Perth and titled "Roses in the Sand."

To get the picture, the team got up in time to pick roses before 4 a.m. at the home of Mrs. Molly Healy. Then off to the beach, where they taped the roses on to rose bushes stuck rakishly in the sand.

Actual photographing began at 5 a.m. to take advantage of the soft dawn light.

"We are grateful to Perth, the sand, the rose plants, and particularly to Mrs. Healy for the results, which we like tremendously," writes "Seventeen's" fashion director, Mrs. Rosemary McMurtry.

COTTON DRESS typical of the collection which the "Seventeen" fashion team brought from the United States to photograph at Broome. Organdie overskirt, which is buttoned on with Australian mother-of-pearl buttons, transforms it into a youthful party dress.

Page 8 The Australian Women's Weekly – February 25, 1959

Fig. 7 Fashion shoot with Broome's coastal backdrop, Western Australia. *Credit The Australian women's weekly*, 25 February 1959, p.8/ Are Media Pty Limited / aremediasyndication.com.au / and National Library of Australia [NLAref138936]

Fig. 8 Broome landscape. *Credit* Image supplied by Paspaley, as featured in secret life of pearls by National Geographic

5 Conclusion: Australian *Pearlescence*

This chapter has explored the variegating meanings of pearls and seashells throughout history, mainly from the ancient Roman and Middle Eastern to Australian indigenous cultures. It also pointed out how writers observed them from various multidisciplinary points of view. The value of pearl products derives their underwater value not only from their symbolic status but also from "their uniqueness and scarcity.... In recent years, luxury goods manufacturers through limited series offers and selective distribution have enhanced the notion of scarcity. Thus scarcity arises from artificial as well as natural production constraints" [23, p. 79]. The mother-of-pearl trade, the luminescent matter that forms on the inside of an oyster pearl shell, ought to be equally be prized accordingly. Moreover, shells returned to the ocean are somewhat considered wasteful. A market for using them in other goods for processing into buttons emerged (as well as for constructing roads and footpaths in some countries). The viability of shell waste and their remnant coastal-scapes/sea-scapes is therefore a luxury-waste product used for paving and road construction.

In the case of Broome pearls, they were discussed as being sustainable and artisanal as naturally sourced pearls as well as viable pearl shells. Indigenous Australian practices for crafting pearl shell and worn as a form of body jewellery has important cultural significance and so the pear shell ought to be equally viable as much as the natural pearl. Changes in the Australian post-war era posed new forms of real scenery that exposed Western Australia for future fashion shoots. The pearl marketing trade

Fig. 9 Kailis pearls. *Photo* A Condello 2020

affected how to sell Broome's sea gems through the tourism industry. Subsequently, it developed into a sustainable luxury venture to attract consumers to purchase coloured sea pearls in the twenty-first century. Broome pearls are thus considered exclusive; you never quite know what you will find.

Broome's pearling beds have provided an alternative solution for pearl fishery farms. The viability of the pearl industry demonstrates how the natural jewel has contributed to the meaning of sustainable luxury, transforming how to measure coastal remnants in the Kimberley region. In addition, this measuring system could include how the pearl rating might be adapted to conserve Broome's existing building-shells. The pearling industry, shell waste and its recycling into viable products, such as infilling remnant landscapes, offer alternative coastal sites to occupy and expand sustainable luxury's progressive impact. Marine-cultural heritage experiences constitute what is a form of Australian *pearlescence* as a sustainable luxury injected with authenticity, reliability and organic substance.

Today, indigenous Australians who reside in the Kimberley coastal areas continue to pass on their pearl shell heritage to future generations. And the global demand for Broome's farming of south sea pearls, as with Paspaley, Kailis and other companies, has formed other authentic business ventures for the pearling industry, since sustainable luxury and jewellery is all about authenticity and where the objects are sourced they do not impinge on the environment.

The circulation of pearls in both the fisheries and the jewellery markets continues to be profitable. Sustainable mariculture exists for environmental benefits for nurturing higher quality of fish in its natural habitat. Shells lie on the seafloor. Perhaps they may offer another form of a circular economy in the ocean as more resourceful underwater luxury goods. Manipulation of shellfish, a selfish and unsustainable problem?

Incidentally, in the mid-1990s, sustainable jeweller Manon van Kouswijk produced a sculpture entitled "soap", entrapping a string of natural pearls inside a bar of glycerine soap, resembling a fossilized witchetty grub inside a piece of iridescent opal. It could be interpreted as a kind of two-edged stroke-of-luck between sustainability luxury and hidden jewellery. Washing hands with a bar of soap, the suds wash away and once after getting closer to the source after many endeavours one reaches for the hidden string of pearls. "A connection could be made here between having to use water to get to the necklace and having to dive through the water to gather the pearls from which the necklace is made" [21, p. 89].

References

1. A pearl that saved a town (1938) The New York Times, 28 August, p 18
2. Bettoni B (2014) Fashion, tradition and innovation in button manufacturing in early modern Italy. Technol Culture 55:675–710
3. Buttons…(1932) Smith's Weeky, Sydney, August 20, p 20
4. By the dawn's early light…' Americas's 4am stint to get some pearly fashion pictures at Broome (1959) The Australian Women's Weekly, 25 February 1959, pp 8–9
5. Cappellieri A, Tenuta L, Testa S (2020) Jewellery between product and experience: luxury in the twentieth-first century in Sustainable luxury and craftsmanship. Springer, MA Gardetti and I Coste-Maniere
6. Clifford P, Semeniuk V (2019) Sedimentary processes, stratigraphic sequences and middens: the link between archaeology and geoheritage—a case study from the Quaternary of the Broome region, Western Australia. Aust J Earth Sci 66(6):955–972. https://doi.org/10.1080/08120099.2018.1563826
7. Condello A (2018) Opal entrepreneurship: indigenous integration of sustainable luxury in Coober Pedy. In: Gardetti MA, Muthu S (eds) Sustainable luxury, entrepreneurship, and innovation. Springer, Singapore
8. Condello A (2014) The architecture of luxury. Ashgate, Farnham
9. DeJean J (2005) The essence of style: how the French invented high fashion, fine food, chic cafes, style, sophistication, and glamour. Free Press, New York
10. De La Bruyere J, Rowe N (1776) The works of monsieur De La Bruyere: to which is added the characters of Theophrastus. Also the manner of living with great men, vol 1, 6th edn. J Bell, London
11. Dirlam DM, Misiorowski BE, Thomas SA (1985) Pearl fashion through the ages. In: Gems and gemology, Summer issue, pp 63–78 https://www.gia.edu/doc/Peral-Fashion-Through-the-Agesv.pdf. Accessed 27 May 2020
12. Edwards T, Sarah Yu (2018) Lustre: pearling and Australia. Western Australian Museum Perth, Scott Print

13. Ehrman E (ed) (2018) Fashioned from nature. V&A Publishing, London
14. Fowler C (2020) Sat. 2 May 2020 Australian pearl breeding success offers hope for industry hit hard by coronavirus. WA Country Hour https://www.abc.net.com. Accessed 27 May 2020
15. Gabet O (ed) (2019) 10,000 years of luxury. KAPH Books, Louvre Abu Dhabi
16. Japs. Sell canned oysters to relieve cheap pearl glut (1937) Daily News, Perth, WA, Friday 12 March, p 2
17. Hightower, Victoria Penziner (2013) Pearls and the Southern Persian/Arabian Gulf: A Lesson in Sustainability," Environmental History 18, January: 44–59
18. Knol R (2016) Galalith: alchemy in the age of plastic, November 30 https://blogs.getty.edu/iris/galalith-alchemy-in-the-age-of-plastic/. Accessed 23 May 2020
19. Kwaymullina S (2001) "For marbles" Aboriginal people in the early pearling industry of the North-West. Stud W Aust Hist (22):53–61
20. Machado P, Mullins S, Christensen J (2020) Pearls, people and power: pearling and Indian Ocean worlds. Ohio University Press
21. Manheim J (2009) Sustainable jewellery. A&C Black Visual Arts, London
22. Moore R (1994) The management of the Western Australian pearling industry, 1860 to the 1930s. Aust Assoc Maritime Hist 16(2):121–138
23. Nash J, Ginger C, Cartier L (2016) The sustainable luxury contradiction: evidence from a consumer study of marine-cultured pearl jewellery. J Corp Citizensh 63(September):73–95
24. Ostenberg I (2009) Staging the world: spoils, captives and representations in the Roman Triumphal procession. Oxford University Press, Oxford, New York
25. Paspaley (2020) Social responsibility. https://www.paspaley.com. Accessed 27 May 2020
26. Paterson A, Veth P (2020) The point of pearling: colonial pearl fisheries and the historical translocation of Aboriginal and Asian workers in Australia's Northwest. J Anthropol Archaeol 57:1–23
27. Pitiot C (2019) Luxury, the sweetness of life. In: Gabet O (ed) 10,000 years of luxury. KAPH Books, Louvre Abu Dhabi
28. Pointon M (2009) Brilliant effects; a cultural history of gems and jewellery. Yale University Press, New Haven and London
29. Reid A (2013) Crossing the great divide: Australia and eastern Indonesia in Macassan history and heritage: journeys, encounters and influences. ANU E Press, Canberra
30. Schorle K (2015) Pearls, power and profit: mercantile networks and economic considerations of the pearl trade in the Romain Empire. In: Romanis FD (ed) Across the ocean: Nine essays on indo-mediterranean trade. BRILL
31. Sempels C (2012) Sustainable development in the luxury industry: beyond the apparent oxymoron. In Hoffman J, Coste-Maniere I (eds) Luxury strategy in action. Palgrave MacMillan, London, pp 160–81
32. Southgate P, John L (2008) The pearl oyster. Elsevier
33. Stephens S (2012) Sex and Spleen: fetish in Baudelaire's 'Les Bijoux. S Central Rev 29(3): 63–79, 199
34. Stott R (2004) Oyster. Reaktion Books, London
35. Szabo K, Koppel B, Moore M, Young I, Tighe M, Morwood M (2015) The Brremangurey pearl: A 2000 year old archaeological find from the coastal Kimberley, Western Australia. Aust Archaeol 80(1):112–115
36. The shells of Western Australia (1961) 29 March, p 58
37. Turunen LLM (2018) Interpretations of luxury. Palgrave Macmillan
38. Travellers and traders in the Indian Ocean world catalogue (2016) Western Australian Museum
39. White R (2018) How the Victorian Craze for conchology became a billion-dollar business, 28 Feb. https://thecabinetofcuriosity.net/2018/02/28/how-the-victorian-craze-for-conchology-became-a-billion-dollar-business/. Accessed 08 June 2020
40. Wiesling L (2019) A philosophy of luxury. Routledge, London and New York
41. Winchester S (1987) An Australian edition of cannery row. The New York Times, 9 August, XX15

The Pearls and Their Soul

Justine Ducrocq, Marion Fossati, William de Marsangy, and Ivan Coste Manière

Abstract Pearls... Long before Fitzgerald's symbolism, Pearls have been building one of the most inspiring treasures, and one of the very first money. Our friends, William de Marsangy and Trésor Noir, the iconic jewellers from Mauritius, are redefining the core activity and meanings of this incredible überluxury handcraftsmanship.

From history to sales, this luxury heaven might be considered one of the best-preserved and protected niche markets. William has been teaching all of us many of his hidden secrets, sharing history in the various pearls production regions of the world, highlighting the transition from fishing to culture, the particularities of their trade and their socio-economic importance. Symbols are everlasting and sustainability is obvious everywhere.

In addition to the commercial aspect, this chapter will be dealing with the sustainable biological, geographical, economic, social and cultural aspects of the world's pearl history over the past.

Pearls… who has not been admiring them, coveting them, or at least dreaming of them? Pearls are the core motivation of William de Marsangy and Trésor Noir and have been the central essential goal of his smashing work for more than 15 years.

As you can discover in the appendix, William has worked on their history in the various pearl production regions of the world, highlighting the transition from fishing to culture, the particularities of their trade and their socio-economic importance.

In addition to the commercial aspect, this chapter will be dealing with the sustainable biological, geographical, economic, social and cultural aspects of the world's pearl history over the past.

J. Ducrocq · M. Fossati · W. de Marsangy · I. C. Manière (✉)
Skema Business School, Université Côte d'Azur Nice, Valbonne, France
e-mail: ivan.costemaniere@skema.edu

W. de Marsangy
Tresor Noir, Grand Baie, Mauritius

I. Coste-Maniere and M. Á. Gardetti (eds.), *Sustainable Luxury and Jewelry*,
Environmental Footprints and Eco-design of Products and Processes,
https://doi.org/10.1007/978-981-16-2454-4_5

It is important to note that not all this work could exist without pearl oyster farming. If we take the example of French Polynesia, the safeguarding and propagation of endangered pearl oysters, pearl farming has turned the Polynesian offer from mother-of-pearl to cultured pearls. Moreover, this shift became indispensable because of the replacement of mother-of-pearl by its equivalent plastic and chemical products ersätze and substitutes.

Nevertheless, pearl farming in this region uses a much smaller area than the one formerly affected by the pearl oyster fishery. Pearl farms were only established in a few Tuamotu atolls, and more recently in the Cook Islands, although research and experiments to develop nacre and pearl culture techniques were underway on several other islands in the region.

This chapter aims, on the one hand, to present the pearls quality approach and typology and, on the other hand, to describe the bio-ecological characteristics of the different species of pearl oysters, thanks to the advances and sustainable innovations that have made their cultivation possible. Besides, the different myths associated with pearls since their discovery will be highlighted.

This research is conceived keeping in mind and highlighting the sustainable development alternatives within the pearl regions of production and farming in the world.

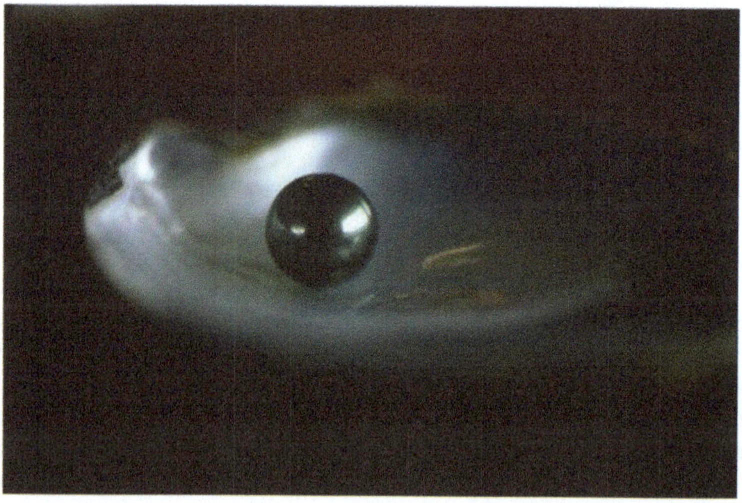

"For the last 15 years I have a been a pearl hunter, between Tahiti, Australia, Dubai, New Zealand, Japan, France, in their paradisiacal environment", explains William de Marsangy, who became a pearl lover and a pearl expert and graduated from the Gemological Institute of America. He, with his wife, is the founder and CEO of Tresor Noir.

Thanks to his past experiences in traveling, he has been introduced to spirituality and aesthetics of the Western and Indigene communities in his work pretty early, which enabled him to have an open mind and feel "in communion with the universe" as he explains.

William's goal is "to inform people of knowing a bit more about this extraordinary gem because it is a subject little or not enough known by professionals and even less by the general public".

The mollusks we meet in our rivers and oceans are our most distant ancestors. At the dawn of time, when vertebrates had not yet appeared, the seas of the globe were already teeming with mollusks. Oysters and shells would one day become man's first delights.

Producing pearls in order to magnify the first jewels was more than obvious, and they even served as the first coins and good luck charms.

Their origin is lost in the mists of time.

The Romans greatly appreciated the gustatory qualities of oysters. They were the first to cultivate and farm them.

They were known as Great oysters as the philosopher Seneca who was famous for consuming a hundred oysters per week.

At that time, eating oysters also was supposed, among other virtues, to ensure its owner's good fortune and health. In addition to its curative virtues, this seafood is also reputed to be powerful aphrodisiacs.

From Roman times to the present day, their dietary qualities, taste and delicacy have made them a staple on the tables of gourmets.

We will begin by defining what pearls are and which are their different forms and classifications.

A pearl is a hard, lustrous spherical mass, typically white or bluish-grey, formed within the shell of a pearl oyster or other bivalve mollusks and should be considered a highly prized gem.

It is a spherulitic calcareous concretion secreted by the mantle of certain mollusks, in particular the meléagrines (known as pearl oysters), in the sea (salted water pearls), and the lake and rivers (freshwater pearls).

The pearls were named MORVARID in Persian, and MARKARIT in Armenian, which gave rise to their Greek name MARGARITÊS, "poetic", then their Latin name MARGARITA.

This last denomination has been adopted by biologists to qualify certain pearl mollusks, especially the Pinctada Margaritifera—the black-lipped oyster more specifically farmed in Tahiti.

The formation of a pearl is quite simple to understand. In fact, all mother-of-pearl-producing mollusks can secrete natural pearls. The initial condition depends on the accidental penetration of a foreign body (grains of sand, worm larvae, little crab...) into the shell. Some of these foreign bodies may be expelled before the pearl germination.

However, others will remain. The oyster will then neutralize the intruder by surrounding it with a "bag", biochemically synthesizing an aggregate of cells of the mantle that multiply. Like a spider, the foreign body will be stuck during a period depending on its origin. After this period the oyster will give birth to the pearl.

Thus, pearls may be divided into two distinct categories.

First, the **fine pearls**, which are natural concretions secreted accidentally and without any human intervention inside mollusks. They are composed of an organic substance (a scleroprotein called conchiolin—SCLER, from the Greek "hard", like for arteriosclerosis, hardening of the arteries).

It is also important to highlight the fact that the word "pearl" on its own can only be applied and used for natural pearls.

Secondly, the **cultured pearls**, which are pearl concretions secreted inside mollusks.

The outer part of cultured pearls consists of organic layers: mother-of-pearl (the so-called conchiolin).

It is made of calcium carbonate (usually in the form of aragonite). The secretion of the layers of mother-of-pearl is a metabolic reaction of living mollusks.

Various human interventions trigger this reaction.

This classification applies to all cultured pearls, whether they have a solid core or are the result of an organic implant.

Furthermore, pearls are mainly produced by the pinctada, or meléagrine (pearl oyster).

As explained before, they grow when they are facing and bothered by an artificially introduced grain of sand, a parasite or a fragment of mother-of-pearl, which is covered over the years within concentric layers of secretion based on aragonite.

The **value of a pearl** depends on many different characteristics:

- Its weight, which is calculated in grains, carats or momme.
- Its color, which can be white, pink, golden, grey or black. Pearls have body colors and undertones. These criteria are important for natural pearls but not for cultured ones since they can be dyed.
- Its regularity and shape, which can be circular, pear-shaped, baroque etc. Usually the rounder it is, the most expensive it becomes.
- Its brilliance.
- Its luster and "orient", which constitute the "soul" of the pearl. It represents this inner glow of the pearl, which justifies its high quality. Orient is the faint lay of colors on the surface of the pearl.

After defining what a pearl is, we will now look at its origins and its past.

From ancient times, pearls enjoyed great prestige. The oldest pearl necklace that we know has been discovered in the tomb of a Persian princess, buried near the ancient city of Susa, and could be dated 420 BC.

The Persian Gulf and the Red Sea have been the major producers of fine pearls since antiquity. For thousands of years, divers there retrieved pearls. We also found evidence that pearls were used during the Han Dynasty by the Chinese divers, using ropes in order to be safely brought back up to the surface. This suggests the existence of pearls back to a prehistoric origin in this region.

Moreover, we have found that the production of fine pearls has also been organized by diving off Venezuela—Ilsa Margarita—in the gulfs of California and Panama, in the north of Australia and in the Tuamotu—French Polynesia.

In Europe, we can easily find a large number of pearls in the Vologne valley in Lorraine, in central Germany, in Sweden, as well as in Scotland. In the United States, various pearls can be found, both in the valleys of the Mississippi and its tributaries.

Chinese royalties were used to receive pearls as gifts as early as 2300 BC, while in ancient Rome, pearls were considered as the ultimate social status symbol. The only persons who were allowed to wear pearls were always members of the ruling class. Therefore, pearls became gradually a means to trade. For instance, in China, pearls were even used to pay taxes.

"Spanning from antiquity to the modern era, it sets out six historical categories of luxury - and relates these to the built and unbuilt environment, taking different cultural contexts and historical periods into consideration"—Annette Condello.

However, Cleopatra has written the most famous legend about pearls, as she crushed a pearl into a glass of wine to prove to Marc Antony that she could offer the best expensive banquet in history.

Pliny the Elder recounts this famous episode in the life of the extravagant Cleopatra:

"At the time when Antony daily feasted on selected dishes, Cleopatra, with the haughty and provocative disdain of a crowned courtesan, denigrated all the sumptuousness of these delicacies. He asked her what could be added to the magnificence of her table; she replied that in a single dinner she would gobble up ten million sesterces. Antony was eager to learn how, without believing it possible."

So, they made a bet; the next day, the day the decision was made, she served Antony a dinner that was not only sumptuous (it was not to be a wasted day) but ordinary. Antoine laughed and asked for an account of the expenses. It was, she assured him, only one on the side; the dinner would cost the set price and only she would eat the ten million sesterces. She ordered the second service to be brought in. As she instructed, the servants placed before her only a vase filled with vinegar whose violent acidity dissolved the pearls. She wore to her ears the extraordinary jewels, this truly unique masterpiece of nature. While Anthony was wondering what she was going to do, she took one of the pearls off, dipped it in the liquid and, when it was dissolved, swallowed it. She prepared to swallow the other one in the same way; L. Plancus, an arbitrator of this bet, put a stop to it and pronounced that Anthony was the defeated one, an omen that was fulfilled.

Fame is not lacking to the twin pearl; after the capture of this queen, victorious in such a considerable debate, she was sawn off so that half of this great dinner could be forging two earrings for the Venus of the Pantheon in Rome (4).

With such a long and ancient past, it is no wonder that pearls have been entangled in myth and legend over time, and still generate so much admiration.

The general tune and symbolic content of the pearl have evolved until modern times.

As he came back to Japan in 1902, Tatsubei Mise, experimented with new studies in pearl farming. In 1905, he announced his success. Nishikawa married the second daughter of Kokichi Mikimoto, a businessman who specialized in the trade of half-cultured pearls, and therefore crowned "King of Pearls" in an article in the New York Herald on October 9, 1904. A real race and war for patents erupted, with disputes between T. Nishikawa and Kokichi Mikimoto, who had also been a pearl farmer for many years, leading him to start his own experiments.

On the shores of Kobe, the island now known as Pearl Island, and hosting a cultured pearls museum where Mikimoto, Nishikawa and Mise are jointly celebrated, Mikimoto developed the technique of breeding in baskets suspended to ropes and used a method for the precise introduction of mantle grafts allowing the secretion of pearl layers around the mother-of-pearl core kernel.

This technique was inspired by the work of his late son-in-law. The dentist Otikichi Kuwabara, a Mikimoto's friend, carried out the implementation.

The secrets of the pearl bedding were well kept, despite the fact that the Americans could not invest in a pearl farm without the presence of Japanese grafters. (Source: magazine "Ze Bijoutier"). Mikimoto opened a jewelry workshop to mount his pearls into jewelry. In the meantime, the world production of mother-of-pearl buttons for clothing began to decline following the massive arrival of plastic shirt buttons.

Then in the 1970s, Tahiti succeeded with his first pearl transplants in the region.

Indonesia and the Philippines also started producing pearls. The Japanese produced their first oyster fertilizations in laboratories.

When the Queen Elizabeth II of England and her husband visited the Mikimoto Pearl Island, the worldwide recognition for Japanese pearls got an everlasting confirmation.

Princess Diana even became the dream ambassador of pearls.

Pearls became more and more popular and became a feminine ornament that is almost impossible to circumvent.

The Nursery is an aquaculture establishment for breeding broodstock or obtaining young spats. According to "Jewelry News Asia", May 1999 both in Indonesia and Burma, there are not enough wild pearl oysters, and the production of these oysters depends quite a lot on sustainable nurseries. It is only there that the spats have a strong chance to survive. The greatest danger to the spat in a hatchery is contamination by external bacteria that would wipe out most of the spat. This is why absolute cleanliness is necessary.

Everything, every little thing from the saltwater pipes, the filters, the pump system, the farm itself does not allow sand or debris particles larger than 5 microns to pass through.

The water is then taken to new tanks where it is filtered with special ultra-modern filters that retain all particles larger than 1 micron (1 thousandth of a millimeter).

The temperature of the water needs to be kept constant. Only one degree could unbalance a very fragile environmental ecosystem.

The water has to be sterilized and the four types of plankton and algae that form the spat are introduced so that they can develop naturally. Wild adult oysters are used, and through the release of sperm and other eggs, millions of spats are produced and obtained.

After three years of maturation and growth, the spats have finished their transformation. Only at that stage, we can proceed to the graft. The graft is a real high-end definitive surgery.

It has been confirmed that the high mortality of pearl oysters after the graft is the result of infection. It is believed that this high mortality is not attributable to any known bacteria, parasites or fungi. Despite 122 attempts for virus isolation, using 14 types of cell lines, no virus could be isolated and identified. There is no evidence that the mortality is attributable to a virus yet, making this production a real dilemma.

In addition, high mortality was observed in 1998 for common oysters in western Japan, which was attributed to a toxic phytoplankton, Heterocapsa circularisquama, and not to a pathogen.

The graft itself, as a trauma, does not only cause mortality in the farms. In Indonesia, due to the El Nifio effect in 1994, the climatic conditions and temperature changed that caused great damages to the ocean. A very high rate of oyster mortality could be observed while the wild oyster stock had already been strongly decreasing because of the excessive fishing practices.

Therefore, the sustainable policy of the government has been introducing temporary laws in order to prohibit all kinds of oyster fishing in order to restore the population. Thanks to this, it only took a few years for a quick return to normal.

All oyster productions depend solely on hatcheries. For the time being, these laws have been shifting the diameter limitation of the produced pearls from 8 to 11 mm.

One of the peculiarities of the country is its equatorial situation, enhancing the stability of the water temperature and allowing the oysters to grow faster. The pearl production is such that the pearl layer produced in 18 months is equivalent to that produced in 24 months in Australia.

The climatic conditions in Polynesia are equivalent to those in Indonesia.

After all these explanations, regarding the history and technical side, let us look at all the evolving symbols that the pearls represent and embody.

Pearls were mystical and moral symbols in various saws because of their mystical and spiritual origin. Sometimes they embodied sacred beings or principles or abstract ideas. From prehistoric times to modern times, the pearl has always been the symbol of perfection and pure beauty, love in fullness, innocence, humility and a sense of gratitude or appreciation.

Hinduism used the image of pearls within its sacred texts as early as 1000 BC. Chinese Buddhism is also including pearls among the eight jewels, also known as the eight precious objects in its iconography. The mysterious cults of Greco-Roman times, which reached their zenith in the third century, also found a special meaning to beads. Early Christians considered these gems as a powerful symbol of purity, faith and the sacred birth of Christ.

The wealth of India and Persia based on pearls has been pushing their inhabitants to appreciate and collect large quantities of pearls. In the East, they appear in religious cults and poetry. In the Veda (1000 BC), the Ramayana and the Mahabharata, there are numerous allusions to pearls. An ancient Hindu myth reports that the elements themselves made these offerings to the deity. The air offered him the rainbow, the fire the lightning, the earth a ruby and the sea a pearl. The rainbow formed a halo around the God, the lightning was his lamp, the ruby adorned his forehead and he put the pearl on his heart.

In the West, the birth of the pearl is associated with Venus; the Goddess of beauty. This legend was represented by Botticelli (1486) in his painting The Birth of Venus, exhibited at the Uffizi Gallery in Florence. Venus is depicted while emerging from a large shell. Pearls are also associated with tears. The white pearls are women's tears; the black pearls are men's tears.

In the religious world, pearls have always been praised as a symbol of purity, humility and fear of God. The first Christians transformed an ancient myth of divine origin into a metaphor of Christ's birth, making the pearl an emblem of both Christ and the Virgin. The pearl also represented the soul incarnated in an earthly body. The pearl soul was innocent, pure and full of faith and wisdom, though surrounded by the corruption of the world. Then the sacred imagery of the pearl focused on Saint Margaret of Antioch, whose name comes from the Greek word for pearl. The pearly gates, also symbolized the entry to the kingdom of heaven, recalling the pearl gate that Aphrodite had. This gate represented the entrance to her sexual paradise.

The association of pearls with sexual pleasure in paradise spread to Islam. Islamic iconography used the pearl as a symbol of divine sexual fulfillment. When the believer died, he was set in a pearl or surrounded by pearls among which he would live forever accompanied by a beautiful woman.

Until the seventeenth century and even long after, Europeans accepted the divine origin of pearls. In Victorian times, as a representation of love, the pearl was a symbol of the clitoris, reminiscent of Aphrodite's pearl door. Its use in religious imagery made it a symbol of salvation, while its monetary value made it a symbol of wealth. Even today, in psychology, the pearl is still used in the interpretation of dreams. It symbolizes the mystical center or soul or the sublimation of abnormal impulses.

Since pearls have enjoyed for so many centuries the status of mystical creation and bearers of supernatural powers, man's obsession with them has not been due solely to lust and wealth, but rather to spiritual or superstitious motivations.

Pearls also have their importance in medicine.

The most common belief was that they had the power to give great vitality to those who possessed it. The association of pearls with the fundamental forces of nature led to the belief that they were an indicator of the health of their wearer. They turned pale if he was ill and lost their luster if he died.

In Babylonia, pearls and the shells containing them were believed to have regenerative powers. The Chinese attributed vital powers to them and the power to prevent forest fires.

The metaphorical and mythological value of pearls made them the subject of study by alchemists and astrologers, as well as dream interpreters. Associated in ancient mythology with the moon, the sun and the sky, pearls have had several meanings for astrologers. Alchemists found many medical uses for them. Modern practitioners of Ayurvedic medicine continue to use some of the alchemists' ancient formulas to promote longevity, virility, good eyesight and digestion. Pearls were also known as aphrodisiacs and were a first-class product for cosmetology in China and Egypt.

The medicinal use of pearls is very ancient. It is mentioned in the Charaka-Samhita, the oldest medical text written in Sanskrit, which dates back to the beginning of the Christian era. It is especially in the East that therapeutic virtues have been attributed to pearls. They were used to treat fevers, indigestion and hemorrhages and were attributed to stimulating virtues.

The literature abounds in accounts of methods to use them as pills, ointments, etc. According to Narahari, a physician in Kashmir in 1240, the pearl cured eye diseases. It is an antidote against poisons. It cured tuberculosis, strengthened and improved general health.

For arable doctors, pearl powder cured sick eyes. It was effective in treating palpitations, nerves, melancholy and hemorrhages. In Japan, even in the twentieth century, pearls were given medical qualities. In China, they were used to treat the eyes and ears. In Japan, they were used to treat insomnia, gynecological disorders, coughs and other ailments. In all oriental civilizations, whole, powdered or burnt pearls were used to treat heart disease, indigestion and to make all kinds of fortifying agents. In Greece, pearls were used as medicines even after Aristotle, Plato and Hippocrates who developed the basis of modern medicine.

Many medieval writers cite the beneficial effects of pearls. Their qualities were so powerful that it was enough to wear them to have their effect believed.

Pearls had a central place in world pharmacy until the advent of scientific medicine in the nineteenth century. It was only then that Western doctors considered pearls to have no more medical value than common chalk. In India, healers use many ancient prescriptions, and in China, they are used in traditional cures.

Since the links between supernatural attributes, medical care and superstition are difficult to discern, it is not surprising that superstitious uses of beads have proliferated. In the Atharvaveda, which dates back to at least 2400 BC, there is a reference to a pearl and mother-of-pearl amulet used by the Hindus to give long life and prosperity to young Brahmin disciples. For Taoist mystics seeking immortality, pearls were an important ingredient in a formula for perpetuating eternal youth. In the book of Han, an ancient Chinese text, it is said that the pearl is the occult soul of the oyster. In the Middle Age, ladies offered pearls to their knights to wear in tournaments.

Finally, thanks to the link of pearls with the notions of purity and softness, since antiquity they have been a suitable wedding gift. According to an ancient Hindu tradition, it is common in weddings to present and pierce a virgin pearl.

At weddings in the West, pearls are also common. Nevertheless, in many Western countries it is considered a bad omen that the bride wears pearls on her wedding day because they represent the tears she will shed during her married life.

"Today, the definition of luxury has lost some of this clarity. How can we define something that is present at so many levels and in such a diverse range of products? Pure luxury no longer exists. The whole concept of luxury is changing, both anticipating and following the constant shift of consumers' dreams, needs and desire."

References

1. Gemperles. Mythes et symbolisme des perles
2. Gemperles. La légende de la perle noire
3. Gemological Institute of America (2002) Gems & gemology
4. Gemperles. Pêche, exploitation, ostréiculture et gisements des perles de Tahiti
5. The University of California Journal of Agriculture (1916) Univ Calif J Agric 4:119
6. Victoria Fanlay (2010) Cleopatra and the pearls
7. Condello Annette (2014) The architecture of luxury. Ashgate, Farnham
8. Hoffman J, Coste-Maniere I (2012) Luxury strategy in action
9. South Pacific Commission (2000) Pearl Oyster Inf Bull (14)
10. New York Magazine Company (2007) New York, vol 40
11. Tcherkézoff S, Douaire-Marsaudon F (2008) The changing South Pacific: identities and transformations, p 333
12. South Pacific Commission (2008) Fisheries Newsletter

Traceability, Sustainability, and Circularity as Mechanism in the Luxury Jewelry Industry Creating Emotional Added Value

Danielle Keller-Aviram

Abstract In the past, designers and artisans had changed and improved the way people live; these days they are recycling trends with little creativity and imagination (Edelkoort 2016). In this paper, I would like to claim that designers should use technological achievements with much awareness and consciousness for the surrounding in order to influence the future for the better. Observing the ever-decreasing prices of apparel items leads one to ponder, how could it be that such prices are tenable? The linear production known from the industrial revolution must change. In the last decades, the fashion industry became more aware of the damages caused as a result of exploitative production, and increasingly more certifications and governmental regulations are used. Unfortunately, this is not the case with the jewelry industry, which has yet to initiate reform. A new set of jewelry criteria, laws, and certifications are in need as their current absence. As a Master's graduate for sustainability in fashion and design, I was in the midst of my research on the jewelry industry, trying to understand and decipher the complex supply chain of practices and processes—from material to making and beyond. **The raw materials** (precious metals and stones)—where are they sourced? Who is mining them and in which conditions? Which additional materials are being used? What kind of processes being made? What are the outputs and impacts of these processes? **Finalizing the materials**, preparing them for production—which chemicals are used? What are the impacts on the environment, the workers, and the communities that surround these sites? What changes could be made in order to minimize these effects? **Designing and making** the products from handmade to 3D printing—which production methods are more efficient? What are the outputs of each technology? What are the impacts of recycling the materials? Could recycling become on a large scale creating a more circular economy? Following the issues above, I used several research methods: in-depth interviews with knowledgeable experts, designers, production employees, and material suppliers, content analysis of international and private-written reports, advertisement content analysis, and implementing these questions on my own M.A. final project products. Based

The research conducted in collaboration with Wikirate project as part of an M.A. thesis.

D. Keller-Aviram (✉)
Berlin, Germany

I. Coste-Maniere and M. Á. Gardetti (eds.), *Sustainable Luxury and Jewelry*, Environmental Footprints and Eco-design of Products and Processes, https://doi.org/10.1007/978-981-16-2454-4_6

on the gathered knowledge and information in this research, I propose a methodology through which brands and designers could examine sustainable, traceable, and circular practices, and how to implement them. Since many designers and product developers are not even exposed to issues like sustainability as part of their training, these aspects could not be taken into account while planning these products. On the other hand, 80% of the environmental impact of a product is decided on the designer table and this is the main purpose of this methodology. Accumulating this database accessible to the industry helps to achieve this ethical shift, hopefully creating a strong connection between all the links in the luxury jewelry supply chain to achieve a sustainable, traceable, and circular industry.

Keywords Luxury jewelry · Traceability · Circularity · Emotional attachment · Craft · Value

1 Introduction

Anthropocene is the current geological period, the time in which human activity is identified as the dominant source of influence on the environment [29]. In today's reality, the leading mindset about products stresses the importance of getting trendy items to the market as quickly as possible rather than creating durable use. Even luxury goods that were traditionally bequeathed from one generation to the next are currently tossed after short useful life. It appears to be human nature to try out anything that seems possible and even mass-produce it without a thought about the next day, let alone the next generation [63].

In the past few years, there has been a growing awareness of and concern about issues of sustainability in the fashion industry and its exploitive system. Such is not yet the case when it comes to jewelry. In this paper, I investigate the luxury jewelry industry from a broadly critical perspective, one of sustainability. By researching this topic from different angles, one indeed gets a wide view of the industry. Based on this research and as one of its outcomes, a new and innovative framework for this industry's operations as a sustainable player is presented. From producers of materials up to consumers, people around the world today are unaware of the impacts of the mere act of mining, processing minerals, or choosing to buy specific pieces of jewelry. Hopefully, this paper and several other initiatives that are trying to push this industry toward transparency and provenance disclosure will allow this topic to be addressed in ways that include practical and attainable methodologies such as that presented below.

As attested in many reports by NGOs, more profound research on sustainable jewelry is needed—one of the reasons and motivations for this study. Only expanding research in this field may clarify how sustainability can become a core value in the design, production, use phase, and end of life of luxury jewelry. Once this happens, designers and companies will be able to implement methods that disrupt the current production chain on the basis of a broader and deeper database. This paper and the research behind it are predicated on several methodologies: in-depth interviews with knowledgeable experts, designers, suppliers of materials, content analysis of public reports with the help and support of Wikirate (an open research platform that focuses on the collection, analysis, and discussion of corporate environmental, social, and governance [ESG] data). The Wikirate matrix is developed collaboratively researching the 15 biggest luxury jewelry brands. The Wikirate platform allowed to develop questions regarding sustainability issues in the luxury jewelry industry and to ask how companies deal with or solve the attendant problems. To gather information about the questions asked, participants are referred to a Human Rights Watch report, *The Hidden Cost of Jewelry*.

After analyzing the interviews with professionals in the fine jewelry industry, a sense of a more robust support system specific to this industry occurred. This system ideally addresses and analyzes challenges with respect to sustainability issues while offering sustainable and traceable alternatives, along with inserting circularity into the sustainable jewelry equation.

Issues of materiality in jewelry-making history will be presented alongside explaining the uniqueness of luxury jewelry. Social, political, and economic changes that led to a shift in luxury production from prior practices to today's are discussed. A presentation of the supply chain is accompanied by a discussion of current production problems from the standpoint of sustainability and transparency. An introduction is given to several pro-sustainability solutions that are used and their limitations, along with the construction of a new methodology and set of requirements for sustainable luxury jewelry.

The purpose of the methodology is to create awareness and engage designers and brand professionals in an analysis of their supply chain and finding the sustainable practices that best fit them.

2 The Lost Traces of the Origins of Precious Jewelry in the Industry: From Past to Present and Beyond

2.1 The Traditional Definition of Luxury and Luxury Jewelry

In antiquity, luxury meant sinfully excessive self-indulgence [32], in which wealth was used to attain a pleasant lifestyle [49]. In the course of history, this definition softened and lost its judgmental tenor but retained the sense of luxury as something enjoyable or comfortable but not necessary [32]. According to the philosopher Yves Michaud, luxuries signify rarity, cost, change, transformation, expenditure, distinction, excess, and pleasure [45]. In the abstract, luxury represents wealth, sophistication, desirability, and influence; for this reason, it is a powerful social force [56]. Luxury is visible in all aspects of life—from how and where people eat to the way people sit on chairs. Yet it is always time- and place-specific and varies dramatically across cultures and times. Luxury is always defined by its surrounding; it depends on what society assumes to be "beyond" necessity [45].

Part of what makes luxury distinctive is the rare aspect of these products from creation to presence [7], the quality of conception [56], manifested in unique or precious materials, the special craft involved, or splendid experience [7]. A luxury artifact can be rare and unusual because it comes from another time or place [45].

Jewelry items as personal ornaments play an important role in luxury culture because they reflect the wearer's values and beliefs via non-verbal communication [34]. It is probable that humans thought of decorating their bodies before they thought of making anything that could remind us of clothing [22]. Jewelry yields information about hierarchy, prestige, power, [9] social status, political ambitions, financial status, gender, ethnic affiliation, and religious beliefs [30].

2.2 Luxury Materials and Making Along with the History

The best way to appreciate the role of objects is to consider them as symbols that represent humans. It is the self that gives them such dignity, glamour, and refinement [46]. Material culture and materials are tangible syntheses or resource flows. They are critical because they make symbolic production real and provide humans with physical means to form identity and act as social beings and individuals [27]. Design matters because it is an intrinsically humanistic discipline of making, connected to the core of why humans exist [37].

Detailed examples that trace the origins of jewelry materials all the way from Asia [45], and of lapidaries and goldsmiths brought from Greece, imply an interest in and appreciation of the origin of such materials [22]. Monarchs imported special minerals from places that they ruled or passed through while traveling from Egypt, Spain, Britannia to Dalmatia, and stones from the Middle East or pearls from the Red Sea and the Indian Ocean. Peridot or topaz from India and Ethiopia and emeralds from Scythia were much desired. The more remote the origin of the material was, the more desirable it was [45].

Luxury products are also the results of skillful production, education, and innovation accumulated over generations [56]. Materials and making is a human urge for thousands of years, this is why becoming a craft master was a prestigious position. The making process plays with people's reflexes, impulses, and emotional tendencies. The hands, arms, legs, and eyes are the human tools in the technical-making process and a window through it the world is seen and understood. [44]. If part of the personal experience of the world is done through the human organs it is clearer why people in prehistory chose materials from their immediate environment that they felt with their bodies [9].

3 The Transparent Supply Chain of the Jewelry Craft as a Result of Traditional Trade Routes

There is historical evidence of the transparent creation of jewelry—from artifacts originating in precious materials to unique techniques and splendid styles that carry the influence of different places. It is not far-fetched to analyze an old piece of jewelry according to today's supply chain and be able to identify most of the processes along the way.

Goldsmiths were international figures, under Elizabeth I of England, generally involved in the financial affairs of clients and trade. It was they who moved gold and silver from America via Spain and Portugal to the rest of Europe, as well as Indian diamonds, Burmese rubies, sapphires from Ceylon, Colombian emeralds, and Persian-grown pearls [45]. Although the expansion of global trade in the seventeenth century made precious materials more available, it was still a source of much pride to share the origins of the techniques and materials used [55].

4 Loss of Transparency During the Industrial Revolution and as a Colonialist Act

The role and meaning of luxury jewelry have changed, mainly due to advanced production and globalization processes [4]. Today, "materialistic" thinking has fallen into obscurity because people use many objects with which they do not intimately identify, neither make them nor understand their provenance [63].

Several key concepts of "modern luxury" jelled in the eighteenth century [51]; since then, goods were copied over into popular culture, transforming luxury to fashionable goods [45]. Therefore, the producer's reputation, represented by name or label, acquired paramount importance in assessing the value of a product as a luxury [45]. Now industrialized, major luxury brands became high-priced commodities that had nothing in common with traditional luxury. The eradication of luxury was accompanied by the loss of knowledge and skills. Here, as in any revolution, the trappings of an elite are overthrown. The old ways of doing things were persistently and deliberately jettisoned in favor of faster and cheaper ones. Traditional exquisite techniques and deep intellectual reflection were lost forever [56]. Luxury is under threat, while its knowledge, tradition, and skills are disappearing. The industrialization of important luxury brands keeps prices high but leaves only a veneer of what makes luxury [56]. Designers themselves are treated as commodities and replaced as quickly as collections change. Luxury stores that once provided an extraordinary retail experience have lost all distinction. Quondam luxury products that had longer shelf life are now replaced as often as the seasons change [64].

The industrial revolution changed the role of jewelry as a symbol of social rank. This evolution created a market for vast quantities of jewelry at prices that the middle class could afford [15]. Hundreds of different components were produced by machines, an electric gold-plating technique was invented, metal alloys replaced gold and silver, and the production of imitation stones increased in quantity and quality [22].

During colonialism, precious resources were used to wield superior power over local communities in developing lands. By the late nineteenth century, there were already examples of Westerners amassing fortunes by exploring Africa for diamonds and other minerals, using the wealth thus acquired for other luxury ventures [45]. In the 1940s, De Beers coined the iconic advertising slogan, "A Diamond is Forever". The tagline forever changed public attitudes about diamonds being reserved only for the rich and ended up creating a diamond demand by convincing people that no marriage is complete without a diamond ring. The growing demand encouraged even more Western companies and businesspeople to control and exploit local communities in developing countries where these precious minerals were found for their own use [61].

5 The Shift from a Linear and Discreet Jewelry Supply Chain Toward Traceability

See Fig. 1.

6 Sustainability Issues in the Luxury Jewelry Supply Chain

6.1 Social and Environmental Problems

Although people spend billions on jewelry every year, manufacturers cannot guarantee the materials and products they sell are either ethical or sustainable. The jewelry industry has been associated, for many years, with creating problems for underserved communities. For communities that rely on this industry, mining under brutal, abusive conditions and morbidity are common situations [40, 41].

One major problem for gold-mining communities is the lack of economically viable alternative responses to major livelihood challenges. People rely on mining to supply their families with basic necessities, risking their lives, and their children with no assurance of a monthly salary [21]. The use of chemicals in mining processes also has pernicious effects due to exposure through the inhalation of dust, fumes, or vapors [40, 41].

The Responsible Mining Index, introduced in April 2018, assesses the policies and practices of 30 large companies that produce a quarter of mined commodities (gold, copper, coal, etc.) in more than 40 countries. Among these firms, 331 workplace deaths were reported in 2015–2016 [58].

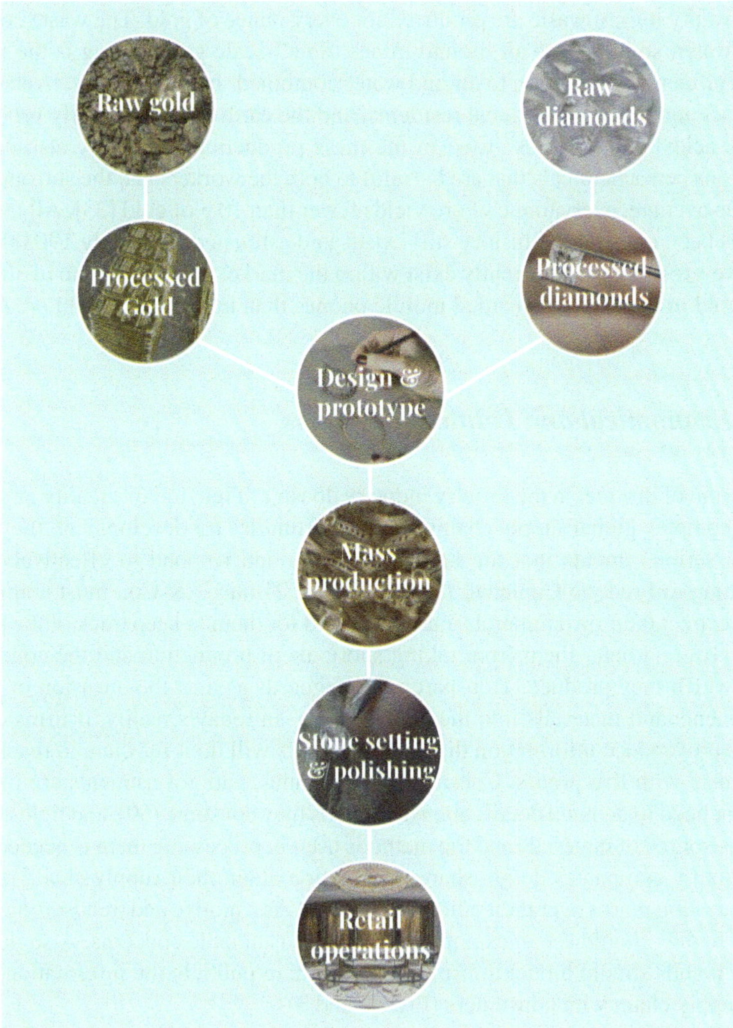

Fig. 1 Supply chain overview (2018); Author's own

The expression "conflict diamonds," first captured the world's attention in the course of an extremely brutal conflict in Sierra Leone in the late 1990s. Rebels against legitimate and internationally recognized governments also used rough diamonds to finance armed conflicts across Africa. The problem here is not the diamond itself as a precious mineral but the people who exploit it to achieve their goals. Diamonds are still used today as leverage over mining communities [8].

Mining for precious materials may result in water pollution, greenhouse gas emission, and soil erosion. Toxic elements such as cyanide and mercury are intensively used in mining; an average large-scale gold mine uses 1,900 tons of cyanide annually

[23]. Twenty tons of waste are produced for every ounce of gold. The waste contaminates water, soil, and the air around mines. Small-scale gold mining is the largest source of mercury pollution to air and water combined. In many mines, oversight is often lacking and workers, local residents, and the earth are not properly protected. Resins, acids, and "pickles" used in the mass production of jewelry also contain dangerous petrochemicals that are harmful to both the workers and the surroundings [21]. On average, every tonne of ore yields fewer than 10 g of gold [33]. All the gold that has been extracted in history still exists, and estimated at roughly 190,000 tons of above-ground stocks currently exist within the market [6]. There is up to 30 times more gold in a tonne of discarded mobile phones than in a tonne of gold ore [54].

6.2 *Economical and Political Problems*

The period of discreet in the jewelry industry doesn't fit into today's reality anymore. While complex global supply chains offer opportunities for development, they often present serious threats that are hard to mitigate and respond to effectively [39]. According to Frederic Cumenal, former CEO of Tiffany's & Co., most companies do not extract their own minerals, making it hard for them to keep track of the provenance [16] and deter them from taking shortcuts in production and the conditions under which they produce. Transparency safeguards against the intrusion of bogus components and materials into the supply chain. In today's reality, if firms do not release provenance information themselves, others will do it for them to reach their consumers with live proofs. Consumers, companies, and governments are increasingly inclined to demand details about the goods they consume [50], and tight control over the source of materials and the methods used in processing them is needed [10]. Most luxury companies do not share information about their supply chain, placing large question marks over their ethical credibility. As a motive and trendsetting force, luxury brands should set an example of solving sustainable concerns. Also, leading luxury brands should model transparency and share publicly the information about their supply chain with consumers (Figs. 2 and 3).

The jewelry industry is highly fragmented, with several large companies interspersed among many family-owned firms that do business worldwide. This diversity and fragmentation present challenges that few industries can match, leading to situations where new materials and models are genuinely uncommon and hard to implement [17].

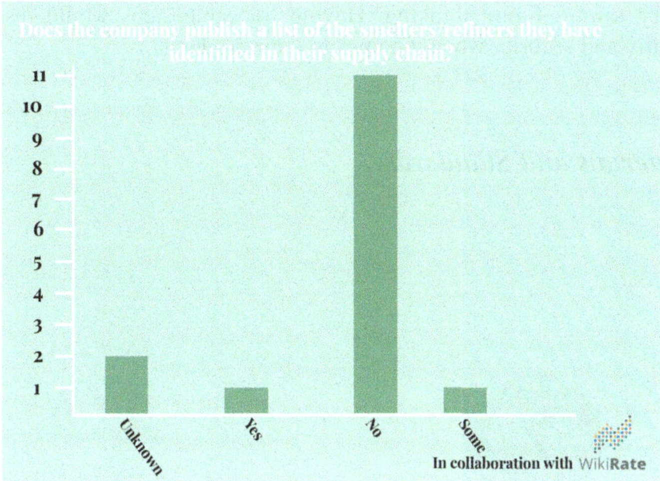

Fig. 2 Minerals origin graph (2018); Author's own in collaboration with Wikirate

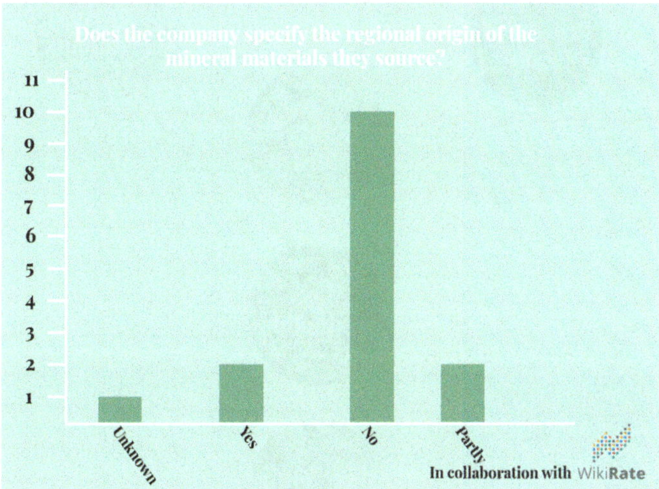

Fig. 3 Disclose refineries graph (2018); Author's own in collaboration with Wikirate

7 New Concepts for Sustainable Luxury Jewelry Production and Supply Chain

Luxury must always stand at the forefront of creativity and innovation in pursuing the highest standards of knowledge and behavior as much as for product quality and refinement [56]. Thus, large companies should use their resources while steering the industry toward future thinking. Having traceable steps would restore their responsibility and security when it comes to their products.

7.1 Materials and Standards

See Fig. 4.

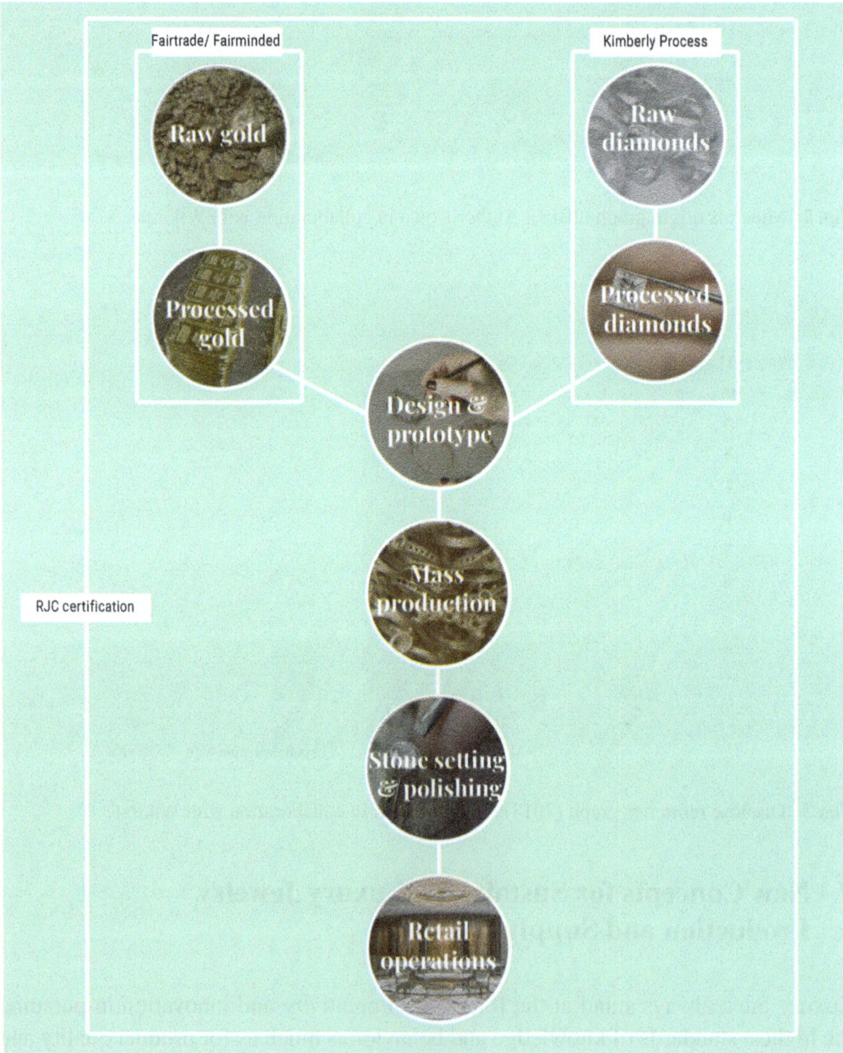

Fig. 4 Standards overview on the supply chain (2018); Author's own

7.1.1 The Kimberly Process Certification Scheme

The Kimberley Process Certification Scheme (KPCS), established in 2000, is a binding agreement among 81 governments, civil society organizations, and industry leaders to stop global trading in conflict diamonds, applying to all diamonds mined since January 1, 2003, incorporating KPCS into domestic law [42]. The scheme has not been updated since its establishment. The Kimberley process is a system of industry self-regulation that requires all traders in rough diamonds, polished diamonds, and jewelry containing diamonds to declare that the stones were from legitimate sources [12].

KPCS refers only to its own definition of conflict-diamond and associated issues and does not oblige members to share information about countries or mines of provenance. Also, any reference to negative social and environmental impacts created by mining and processing diamonds is often disregarded, leaving traceability and transparency off the scope. The certification process is oblivious to money-laundering and mine-rehabilitation issues. In addition, this certification unifies the entire industry, including governments, and social initiatives around the world. The non-inclusion of social and environmental requirements leaves the industry to underperform in sustainable standards. The last issue with KPCS is the laxity of its enforcement mechanism. Although the KPCS organization reports that 99.8% of diamonds traded today are conflict-free, it is almost impossible to verify it.

7.1.2 The Fairtrade Gold Standard and Fairmined Gold—Raw and Processed Gold

The main issues concerning Fairtrade Gold relate to the use of toxic chemicals. While responsible usage and disposal are controlled, the requirements regarding the rehabilitation of closed mines are inadequate. In the present writing, the discussion about rehabilitation is in its initial stages, a complex problem for the future because in absence of requirements or laws constraining companies to deal with their production footprints, closed mines are left as-is.

The Fairmined standard was understood by multi-stakeholders as a robust and credible instrument. This standard focuses on the ability to deliver positive impacts to miners and their communities, a valued tool for the transformation of artisanal and small-scale mining into an active force for good while supporting sustainable-development mining communities [24]. The standard requires minimization of ecological disruption and full traceability to each mine, quite effectively. At the moment, the main problems are a weak extension of the standard to the tier of jewelry additional elements that are used in production and are not available in the market at all, and small quantities of gold, in general, are available in the market (412 kg of Fairmind gold in 2019).

7.1.3 RJC (Responsible Jewelry Council) Certification

The Responsible Jewelry Council was founded in 2004 by a small group of fourteen companies and trade associations interested in making consumers confident in the jewelry they purchase. RJC members commit to and independently audited against the RJC Code of Practices, an international standard on responsible business practices in diamonds, gold, and platinum-group metals. It addresses human rights, labor rights, environmental impacts, mining practices, and product disclosure along the supply chain (Fig. 5).

The research with Wikirate proved that this certification is widely used and known in the industry. The RJC board accommodates only industry representatives, leaving all the other stakeholders out. A new progress report published by the RJC in May 2020 highlights new strategies and actions. During 2019, the RJC launched a new Code of Practices standard, including the expansion of the material scope to include silver and colored gemstones as well. The RJC also entered a partnership with The United Nations Global Compact to launch the Sustainable Development Goals (SDGs) Action Platform—to advance the implementation of the 17 Sustainable Development Goals [60]. Now RJC members recognize only refiners that operate according to OECD-aligned standards as "responsible" for its mineral regulations but until recently, this new revision and regulations were not included in the RJC requirements from brands and businesses. The COP requires companies to go through the human rights due-diligence process that seeks to identify, prevent, mitigate, and account for human rights impacts in line with the UN Guiding Principles. Until this last report, little details were provided on how to carry out and publicly report due diligence in the supply chain or pieces of evidence to ask their suppliers to provide. However, this has changed in the last revision together with aligning their Code of

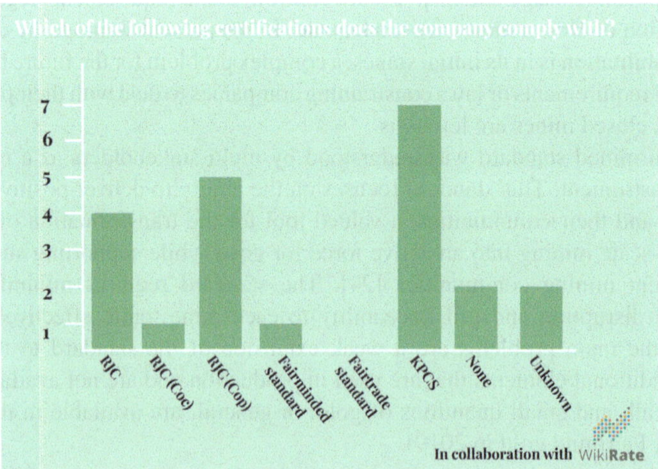

Fig. 5 Certification use among jewelry companies graph (2018); Author's own in collaboration with Wikirate

Practices with the OECD Due Diligence Guidance and the UN Guiding Principles on Business and Human Rights. According to an RJC auditor, there are still some blind spots in this standard. Suppliers only need to pledge that they carry out "strong" human rights due diligence. RJC members are given two full years to comply with the standard after joining; in the interim, they benefit from the RJC's reputation. The same goes for a small subsidiary office with a similar name to the large jewelry firm that can obtain RJC membership and leave the rest of the corporate group out. Still, the entire group enjoys the reputation-associated benefits (Human Rights Watch [40, 41]). The Code of Practices allows companies to withdraw from certification facilities that they do not fully control. The reporting obligations in the code are vague and absolve firms from the need to identify, assess, and mitigate risks in their supply chains.

7.1.4 Alternative Materials, New Approaches, and Models Along the Supply Chain

Currently, replacements for many supporting production processes, like cadmium-free solders and fluoride-free fluxes, are available in the market. The main alternatives include: Replacing "Jewelers' pickle" with citric acid or salt and vinegar, using biodegradable soaps, lead-free enamel, using raw eggs as a replacement to oxidize metals, electrolyte etching, and silica-free polishing compounds.

Establishing closer communication with consumers, encouraging them to customize their demand for luxury products to enhance emotional value and attachment, is a good strategy to deal more responsibly with precious minerals. Retailers should also develop takeback systems through which materials may be recycled or reused, enhancing customer loyalty to the brand. Customers could send old, inherited, or unused jewelry for refinement and remanufactured into new products. Jewelry has the potential to become one of the most sustainable products in the market. Gold especially and metals in general are the ultimate recyclable materials; it can be melted down and turned into different products again. Being recycled and upcycled endlessly is part of the gold product's D.N.A [33]. Finding responsible and environmentally friendly ways of recycling used metals in electronics needs to be developed. There are 50 million tons of e-waste yearly worldwide, and repurposing fine metals from e-waste is an essential part of adopting the circular economy model into the jewelry industry. Although recycling will not solve the problem of scarcity of these precious materials, it certainly would help [52]. The gold industry currently consumes between 4,000 and 4,500 tons of gold in a given year, about half of that comes through jewelry. Theoretically, a situation in which no mined supply is needed, and all the demand can be pulled from what is being held in above-ground stocks is possible if recycling will be widely used and traced [6].

7.2 Technological, Political, and Economical Tools

Today, diamond workshops can laser-inscribe every stone with a microscopic code, indicating its provenance and every step along the supply chain [16]. New generations of tags can be used to label jewelry inconspicuously. Data can be stored and updated as the item moves through the supply chain, creating opportunities for brands to use for continuous improvement. This means that consumers can gather provenance and supply chain information for specific items and companies get reassurance for their goods [50]. Blockchain technology makes the lifecycle of mined diamonds traceable. Attributes of each diamond are recorded digitally, eliminating all risks of fraud [69]. De Beers stated in 2018 that it aimed to create an industry-wide blockchain to track gems each time they change hands from extraction [25]. Lab-grown diamonds offer a viable alternative to dealing with social challenges. Several start-ups have developed such gems using only renewable energy. Traditional diamond giants criticized these products harshly, impugning their quality and purity, but recently all the big diamond-producing companies started producing them. 3D printing technologies that are commonly used in jewelry production have the potential to enable speedy, on-demand, customizable jewelry production minimizing overproduction. This allows companies to speed up their production once there are orders and not have a massive inventory [69].

The European Union has been restricting the use, import, and export of products that contain nickel, given the toxicity of this element and its implication in common allergies and skin problems. The same restrictions and controls should be extended to sustainability problems in jewelry manufacture. EU policy and legislation can ban materials and processes while setting an example for the rest of the world.

8 A New Definition of Sustainable Jewelry and a New Set of Sustainable Requirements

The luxury industry is a leading player in the broader industry, consistently setting new trends and innovations. Thus, leading companies in this industry should aim to combine traditional luxury with new sustainable frameworks that are relevant to today's economy and reality. Minerals and precious metals that retain their underlying social and chemical value should be reused, repaired, remanufactured, and recycled on a fundamental and routine basis. At present writing, no links exist between suppliers along the supply chain, as seen in the circular model presented below. Although the materials are precious, incentives to collect and save them are not offered; therefore, it is not surprising that some even end up in landfills. Since these materials do not decompose in landfill, their composition and value justify a refining on-going, circular process (Figs. 6 and 7).

Apart from integrating circular strategies into production, additional standards in the industry are needed. The KPCS should expand covering sustainable concerns

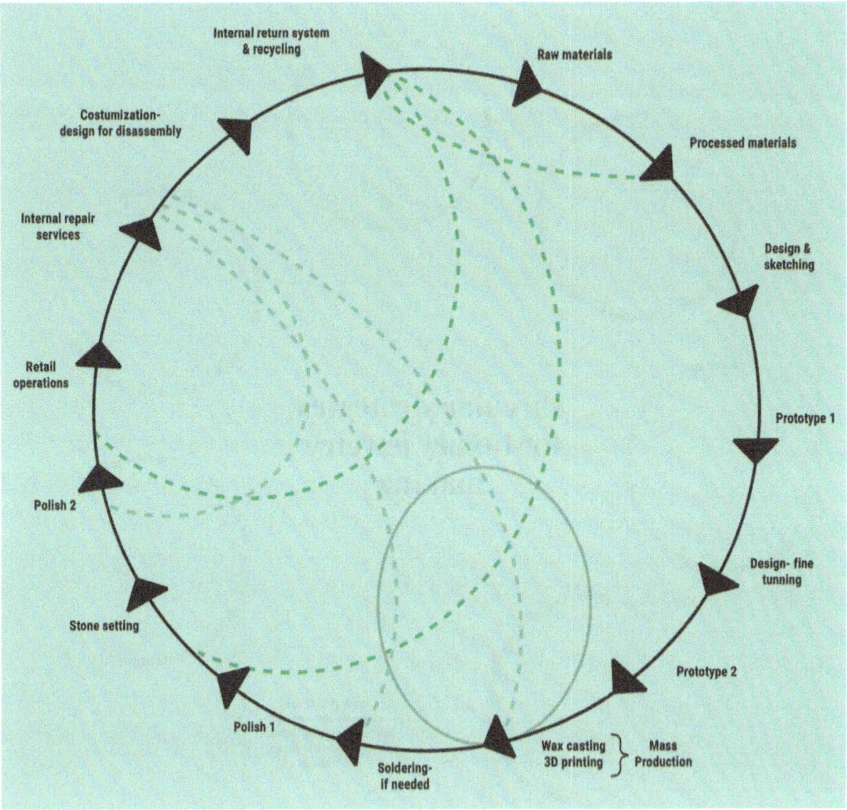

Fig. 6 The jewelry circular model (2018); Author's own

in their broader context, including environmental and broad social issues associated with the diamond supply chain and ground rules for "recycling" diamonds. A similar regime should be developed for precious stones. Improving and restructuring the RJC certification is needed, including the use phase and after-life reference as well as KPIs that measure companies' performances. Last but not least, a standard for recycled gold, including defining "recycled gold" and "recycled gold product," is needed. This would hopefully broaden the use of recycled gold and would create the traceability and reassurance system around recycled content (Fig. 8).

Fig. 7 The jewelry loops (2018); Author's own

9 The "See Through" Methodology as a Tool for a Transparent, Circular, and Sustainable Luxury Jewelry Supply Chain

9.1 Sustainability and Traceability as a Unique Selling-Point Opportunity

Marketers of luxury jewelry base their branding strategies on *legacy* as a powerful value that differentiates their jewelry from mass-market merchandise [1]. Important parts of *legacy* include tradition, appreciation of quality materials, and exquisite production methods [67].

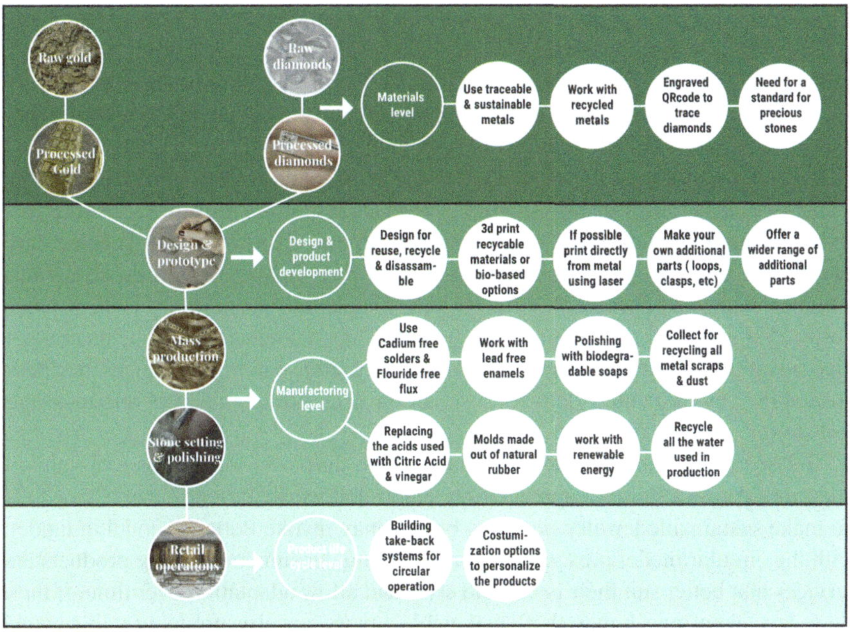

Fig. 8 New requirements for sustainable jewelry (2018); Author's own

To mobilize today's consumers, their emotional buttons must be pushed [26]. One such button is their wish to be responsible global citizens. Some 87% of adults say that sustainability impacts their purchasing decisions. Brands should follow their customers' lead, placing a broad sustainability approach at the heart of their infrastructure and making an authentic commitment [66]. The advertising executive David Ogilvy [53] once stated "The consumer's not a bloody fool; she's your wife." Dana Thomas elaborates "The consumer wants to know that Piaget watches are made in the Piaget factory. [That's] what makes it special. Otherwise, it's just another brand" [64]. Exquisite craft matters less to millennial shoppers, it is more the brand, the image, and the story behind the product that attracts today's consumers [2].

Millennials representing $2.5 trillion in annual spending power are becoming dominant in the labor force. Losing faith in governments and nation-states made them look to other institutions to advance the sustainability goals that they consider essential for a bright socioeconomic future [63]. The contemporary consumer is searching for new luxury products that embrace human values and high ethics [59, 68].

Technically, today all products are "Made In" one single country. This is the current economical and technical structure of supply chains worldwide. In reality, this is never the full story of today's international supply chains and products. Products today are made from many different materials sourced from various countries and assembled together in different locations. There are barely any items in stores that are produced from beginning to end in one country of origin. An entirely new labeling system that would fully answer the questions of who made these products, from which materials, and where is in need. Thus, consumers would be more reassured about what they are buying and from whom [62]. Combining a more value-based perspective with traceability and the circular approach could increase the jewelry industry's uniqueness and differentiation in the market, enhancing profit margins if communicated in the right way to the right audience [47, 48].

Few people can actually tell the difference between an authentic Rolex watch and a fake one. Disclosure of origin will soon be an essential part of establishing trust and securing a reputation for brands and consumers. Companies that become early adopters of transparency and reliability toward their customers will probably have a higher chance of success, a strong brand image, and profitability [50]. Michael Porter [57] analyzed value creation as added-value, a demarche that yields competitive advantages and a unique story. Ultimately, transparency enhances business' profitability by allowing prices to raise and project strong and authentic messages and stories.

Today, jewelry is made on such a large scale that it no longer has real value; it does not express a sense of appreciation, emotional worth, or long-term legacy [11]. To make sustainable jewelry possible, brands may invoke Porter's model in tandem with the circular model presented before, offering consumers ongoing products and services that better suit their needs and style and allow adaptation over time. If these steps are taken, one hopes that sustainable luxury jewelry products will become heirlooms, retained in the long term as the precious objects that they were.

It is the responsibility of brands and designers to make sustainability cool, appealing, and easily communicated [43]. From the perspective of today's economy, corporate social responsibility is an opportunity and not an obligation—a chance for businesses to use their profits, power, and influence to create positive change and empowerment. Brands that adapt to these standards would survive [65]. A new generation is marching toward a revolution; it wants to use products that tell a new story and takes tomorrow into account [31]. High status is increasingly associated with transparency and responsibility, consumers vote with their wallet, and what they want from their vote is value and positive impact [14].

10 The See-Through Methodology—Description, Visualization, and Target Group

As customers take a greater interest in the origins and authenticity of the things they buy, producers and retailers should include provenance tracking tools in the marketing mix [50]. Brands should equip themselves with daily work tools that support their decision-making, especially in a topic as complex as sustainability. By continually measuring and monitoring the entire supply chain, brands can start doing so. This methodology is aimed at creating a reliable industry information platform. A wide-ranging methodology includes supporting and encouraging businesses to approach sustainability and implement it in their daily strategies along their supply

Fig. 9 See-Through methodology (2018); Author's own

chain. The first step for a brand is to investigate its supply chain, including its direct and indirect operations, to understand where the biggest impacts occur and in what places changes should be made [29] (Fig. 9).

Design choices are important for more than a product's final look and function; they also influence its social and environmental impacts, for better or worse. Thus, the more knowledge the designers have, in regard to sustainability issues, the better they can adapt their designs to today's global problems [19]. Changes can also be made closer to the consumer since luxury companies have numerous retail operations worldwide. By integrating services and takeback systems in their retail operations, they can offer products for longer use, introduce customization options, provide recycling incentives, and invoke other circular strategies [38] (Figs. 10 and 11).

In view of the jewelry supply chain presented before, the user of the See-Through methodology moves along in the planning of a jewelry item—from product development and its technical details to suppliers details, locations, certifications (if known), relevant country regulations (if known), and any other relevant details. This process is built on a series of actions that apply sustainability perspectives to the supply chain, from sustainable design strategies, industry-certified materials, and standards, alternative production practices to developing in-house circular strategies in luxury jewelry brands.

One outcome of this methodology is a document summarizing the technical details and offered recommendations and alternative production processes for a planned

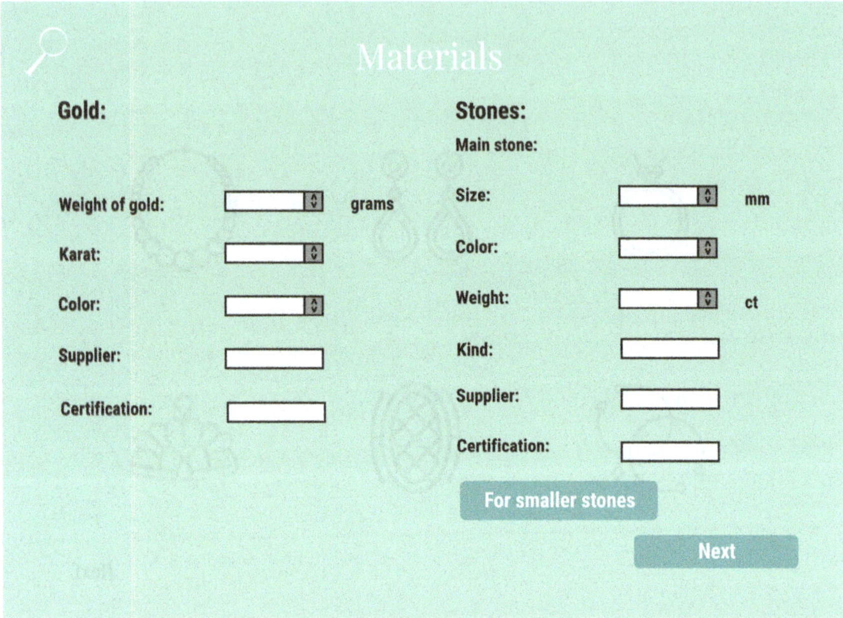

Fig. 10 See-Through methodology (2018); Author's own

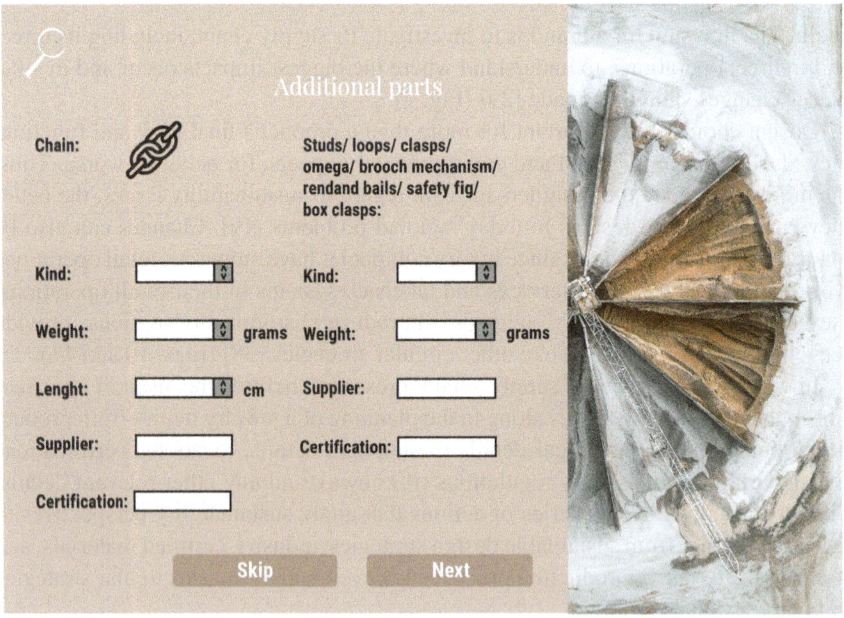

Fig. 11 See-Through methodology (2018); Author's own

item. This document can help the user execute a step-by-step action plan based on recommended sustainable strategies, resulting in a less disruptive industry.

Young professionals are one target group of the See-Through methodology. They may be managers in jewelry companies who are involved in supply chain management or production; corporate social-responsibility or sustainability team members. The other target group is independent designers with their own luxury jewelry brand who deal with all product development, sourcing, buying, and executing in-house.

After analyzing the interviews done for this research with professionals from the jewelry industry, it was clear that dealing with sustainability issues and finding reliable solutions is complex and time-consuming. This might explain why the jewelry industry is far behind, say, the food or fashion industry in sustainable and transparent performances. The See-Through methodology helps to gather and verify the kinds of information mentioned, resulting in the process of decision-making on sustainability matters much shorter and transform these strategies and recommendations into real-life company policies. Luxury is no longer just about exquisite materials, perfect production quality, and superb creativity. It is also the assurance that the brand behaves consciously, lives up to its values, and offers a deeper and more meaningful consumption experience [20] (Figs. 12, 13, 14, and 15).

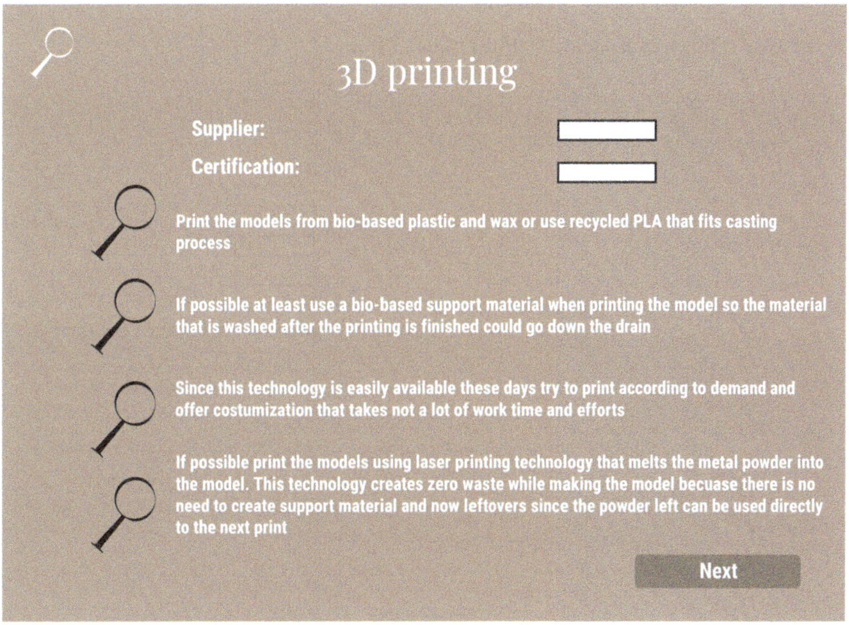

Fig. 12 See-Through methodology (2018); Author's own

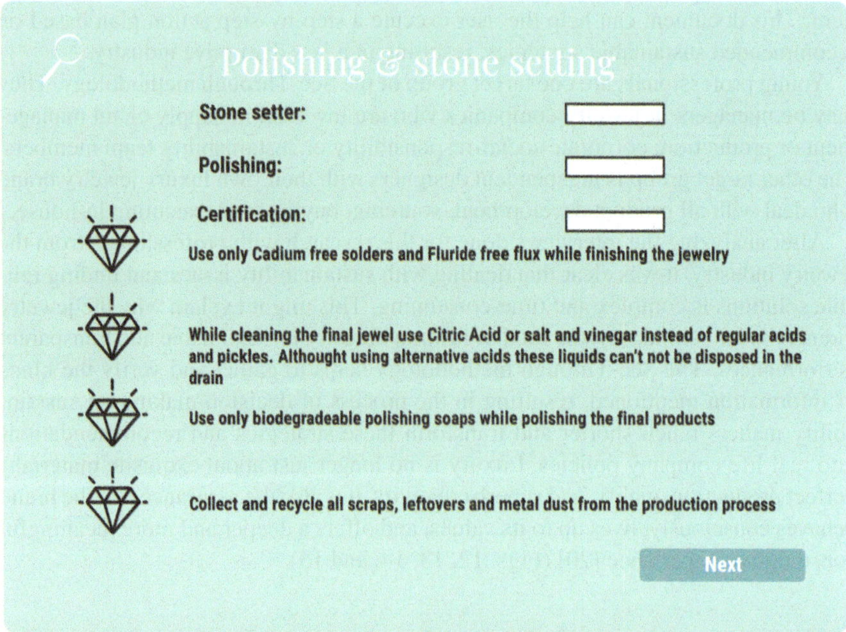

Fig. 13 See-Through methodology- (2018) Author's own

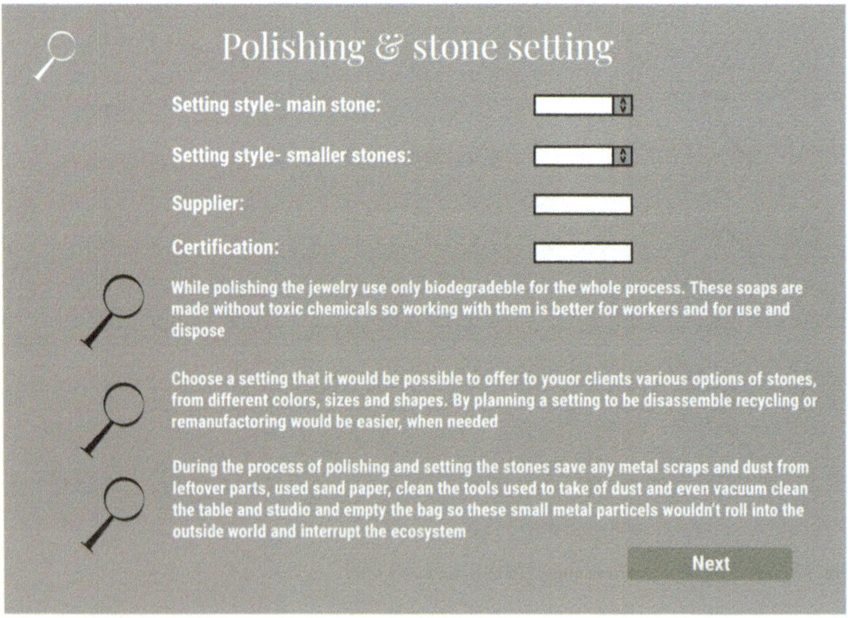

Fig. 14 See-Through methodology (2018); Author's own

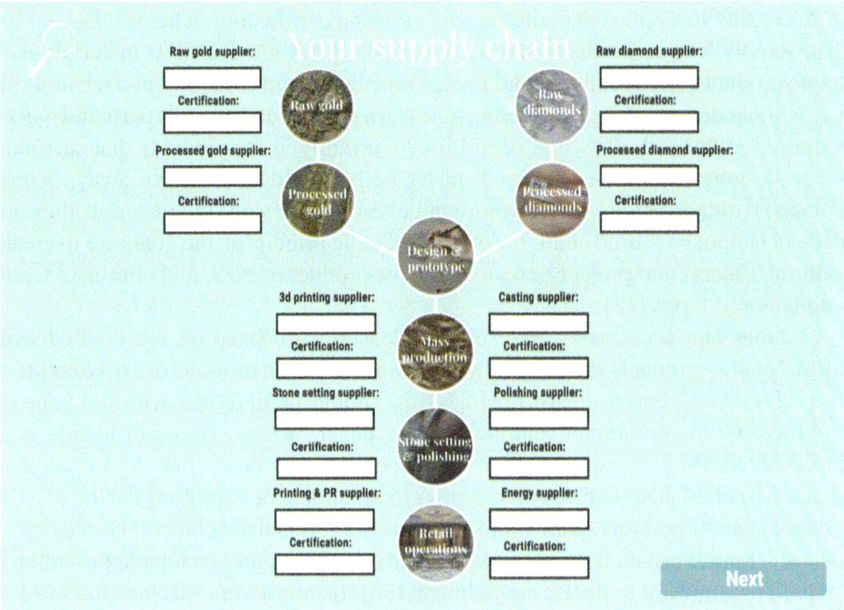

Fig. 15 See-Through methodology—supply chain analysis (2018); Author's own

11 Design Strategies for Transparent and Traceable Luxury Jewelry

Designers today have the power to influence people's behaviors and attitudes toward many products. If this power is appropriately honed, with sustainability in mind, the values that these products deserve can be restored.

Jewelry items are emotive, unique, and expensive. Even as few sectors can claim the status of "belonging to a family," jewelers and jewelry brands are often referred to as the "family jewelers or brands." This status is earned through trust and this prestigious place should be treated with respect.

The See-Through methodology emphasizes a connection and information transfer between the user, the product, and the materials from which the product is made. This may precipitate major systemic changes toward the appreciation and trust users feel for their jewelry. A balanced view of manufacturing includes the notion that thinking and feeling are part of the process. Consistently, today's designers should completely re-think the way they design. When one speaks of the long tradition of jewelry making, a strong link and connection to craft are found, an enduring, basic human impulse: the desire to do a job well for its own sake. It is a quality-driven work. To do good work means being curious about it, investigating it, and learning from ambiguity [63].

According to Cradle to Cradle organization and certification scheme, "Design for Disassembly" means products that are intentionally designed to allow materials used in one product to be reclaimed and reused repeatedly and endlessly in a meaningful way without downcycling. Answering how can a product and all of its parts and pieces be reused at the end of its first useful life? Currently, most products that surround us are designed as though they will never be broken down, thrown away, or just replaced with newer versions for short and one-time use since the materials they are made of cannot be reused at all. In Cradle to Cradle principles, the goals are to create enduring objects and projects, create value for product owners, and eliminate waste within closed loops [13].

Fashions and accessories give sensual pleasure [5]. Since we are flooded with "stuff," quality products should tell a story or an interesting message or evoke emotion [63]. "Products," Farrah Floyd [28] advises, should be designed with the hope of evoking a strong emotional sentiment in the buyer, wearer, who will ideally take impeccable care of them, and who will hopefully pass them forward as precious pieces to be cherished. The aim of designers today should be to prolong the life of each product by incorporating quality and enhancing emotional attachment. For an object to have a long lifespan, it must be aesthetically pleasing and sustainable providing a continuous source of aesthetic nourishment [36]. If *storytelling* was the buzzword in 2016, in 2018 it was *influence through emotion* and hopefully, this would continue to guide designers onwards [66]. Jewelry making is an ancient craft, and as one, people buying jewelry expect them to be unique pieces that have a history and could be inherited. These are part of the DNA of jewelry. This strategy should be revived by telling the full story of materials, workers, craftspersons, and technologies, along with their long journey around the world until they reach the point of purchase [20].

In the report "No customer left behind: How to drive growth by putting personalization at the center of your marketing" by McKinsey, it is counseled that businesses who wish to grow must place personalization at the core of their marketing and design efforts. Today, consumers expect companies to hear their voices and tailor their products and services to their needs [3].

Repair is a neglected and poorly understood but important strategy as well. The sociologist Douglas Harper [35] believes that making and repairing constitute a single whole; makers that practice both possess the "knowledge that allows them to see beyond." This knowledge is the medium that makes making and fixing parts of a continuum. Indeed, it is often by repairing things that we get to understand how they really work [63].

Designers have vast power in today's economic world; they determine the future of our material culture. Almost 80% of a product's environmental impact is decided on the designer's table. Since design originally reacted to problem-solving, today it should take into account the climate crisis, population growth, social aspects like inequality, transparency, and many more sustainability-related challenges. Designers today should not help to produce more of the same; their helpfulness may be expressed in turning out fewer but better products and solving present-day problems [18].

12 Final Remarks

As presented in this paper, the topic of sustainability in the luxury jewelry industry is in the starting blocks and should expand by further research and also by developing suitable frameworks, strategies, and policies side by side to new materials, standards, production methods, business models, and circular strategies. Specific research tailored to this industry would lead to customized solutions. The lack of reliable data on the topic and the dearth of sustainable-materials suppliers, craftspersons, and manufacturers severely impair the industry's ability to move forward and join other industries in the sustainability queue.

Although the historical meaning and understanding of luxury have not changed drastically, the production of luxury has undergone a massive shift. Today's luxury items are not made in Italy, France, or Switzerland, as Dana Thomas expresses in *Deluxe: How luxury lost its luster*. There is a need to learn from the past—from the exquisite technical skills of ancient Rome, the Renaissance style, and the colorful enameling and brilliant cut of diamonds in the seventeenth century. Luxury should benefit consumers not just by feeding their excessive self-indulgence but also by providing emotional connectivity and ethical values.

It is clear that although sustainability is still a niche topic of concern, demand for it is growing and the jewelry industry will have to join, too. Luxury businesses should use this momentum to be at the forefront of creativity and innovation, pursuing the highest standards of knowledge and behavior as much as for product quality and refinement [56] and, by doing so, staying the industry trendsetters. In this industry, due to its structure and complexity, the change should start specifically among large luxury firms that can put pressure on, and make demands of, all tiers. Incorporation into today's economy parts of the traditional definition of luxury products, such as craft, leadership, influence, and change, may restore the supply chain's sparkle. The earlier these brands take a new perspective on these products, the less their reputation will be affected.

The innovative part of this research concerns the establishment of new requirements for sustainable luxury jewelry, and the methodology developed and presented. As the methodology was being developed, no other tool for the production of sustainable, traceable, and circular jewelry was found. That is, neither far-reaching circular requirements nor for the design and product development of sustainable luxury jewelry appears to exist. Based on the interviews conducted, a stronger support system for industry players was requested and this is part of the aim of the methodology. The weakness of this research is that the results were not tested in a large corporate setting to determine its fitness for the realities of this industry and the possibility of adapting and adjusting it to firms' daily operations. Until research in collaboration will take place, it will be hard to say with total confidence that the methodology developed, and the requirements set forth, are suited to the reality of the luxury jewelry industry.

The design has the power to make things look appealing, even within a complex and controversial framework such as sustainability. The jewelry industry should join the sustainability revolution and prove that luxury and magnificence can be sustainable as well. It is a time for design to be produced and judged not only by its optics but also by a more complementary perspective that takes into account the joy and happiness not only of the current user but also of future users of the same product or materials. This, one hopes will lead to a more positive future, in which humankind's impact on the planet will be less negative and more regenerative.

References

1. Au K (2018) Wake up: the time for sustainability in the luxury industry is now. TechNode. https://technode.com/2018/01/18/luxury-sustainability/. Accessed 29 Jun 2018
2. Bain M (2018) Nobody knows what luxury is anymore. Quartzy. https://quartzy.qz.com/1303457/nobody-knows-what-luxury-is-anymore/. Accessed 2 Jul 2018
3. Boudet J, et al (2018) No customer left behind: how to drive growth by putting personalization at the center of your marketing. McKinsey & Company. https://www.mckinsey.com/business-functions/marketing-and-sales/our-insights/no-customer-left-behind?cid=other-eml-alt-mip-mck-oth-1808&hlkid=cfebdd1d03c94f8d8a84653ad0cab1de&hctky=10426092&hdpid=4a80af03-ea92-46a3-a316-5bf53d19f517. Accessed 5 Aug 2018
4. Buchli V (2001) The material culture reader. Berg Publication, Oxford
5. Busch OV (2018) Fashion and biosocial growth. In: Bruggeman D, King I, Pildner K, Smelik A, Teunissen J, Eijnde J (eds) Fashion colloquium: searching for the new luxury. ArtEZ university, Arnhem, p 16
6. Callaway G, Ramsbottom O, London S (2020) Gold in an era of drones, deep mines, and de-dollarization. McKinsey. https://www.mckinsey.com/industries/metals-and-mining/our-insights/gold-in-an-era-of-drones-deep-mines-and-dedollarization. Accessed 15 May 2020
7. Campbell C (1996) The meaning of objects, and the meaning of actions, a critical note on the sociology of consumption and theories of clothing. J Mater Cult 1:93–105
8. Cartier (n.d) Cartier corporate responsibility policy. Cartier. https://www.cartier.com/content/dam/cartier_dam/Communication%20assets/Articles/La_Maison/cartier_politique_corporate_responsibility_en.pdf. Accessed 18 Apr 2018
9. Chadour-Sampson B (nd) History of jewelry in fashion. Fashion history. https://fashion-history.lovetoknow.com/fashion-accessories/history-jewelry-fashion. Accessed 30 Apr 2018
10. Clancy H (2015) State of green business: supply chain transparency ramps up. Greenbiz. https://www.greenbiz.com/article/state-green-business-supply-chain-transparency-ramps. Accessed 31 Apr 2018
11. Clark C (2018) How to make sure your next piece of jewelry is planet friendly. One Million Women. https://www.1millionwomen.com.au/blog/find-out-if-your-jewellery-planet-friendly/. Accessed 31 Aug 2018
12. Conflict Diamonds (nd) Diamond facts. Diamond Facts. http://www.diamondfacts.org/conflict-diamonds/ Accessed 9 May 2018
13. Cradle to Cradle Organization (nd) Company by company, product by product, the cradle to cradle products innovation institute is changing the way we make things. Cradle to Cradle. https://www.c2ccertified.org/. Accessed 10 Jul 2018
14. Craik J, Peirson-Smith A (2018) Redefining and reevaluating new notions of luxury fashion through the lens of conscious consumerism. In: Bruggeman D, King I, Pildner K, Smelik A, Teunissen J, Eijnde J (eds) Fashion colloquium: searching for the new luxury. ArtEZ university, Arnhem, pp 44–45
15. Crowley H, Driscoll-Goulay C, Niemtzov E, Norton T, Prattico E, Woods B (2015) Climate change: implications and strategies for the luxury fashion sector. B.S.R. https://www.bsr.org/reports/BSR_Kering_report_Climate_Change_implications_and_strategies_for_the_luxury_fashion_sector.pdf. Accessed 10 May 2018

16. Cumenal F (2017) Tiffany's CEO on creating a sustainable supply chain. Harvard Business Review. https://hbr.org/2017/03/tiffanys-ceo-on-creating-a-sustainable-supply-chain. Accessed 8 May 2018
17. DeMarco A (2016) Responsible sourcing takes hold in the jewelry industry. Forbes. https://www.forbes.com/sites/anthonydemarco/2016/10/17/responsible-sourcing-takes-hold-in-the-jewelry-industry/#ce03cb21dba1. Accessed 5 May 2018
18. Design Boom (2017) Philippe Starck rejects design art. Designboom. https://www.designboom.com/design/philippe-starck-rejects-design-art/. Accessed 31 Aug 2018
19. Design for Longevity (nd) Design for longevity. Design for longevity. https://designforlongevity.com/. Accessed 4 Jul 2018
20. Doyle I, Bendell J (2011) Uplifting the Earth: the ethical performance of luxury jewelry brands report. Life Worth http://www.lifeworth.com/consult/wpcontent/uploads/2011/06/UpliftingTheEarth.pdf. Accessed 23 Jun 2018
21. Earthbeat Foundation (nd) Für faires Gold. Earthbeat Foundation. http://www.earthbeatfoundation.org/inform/. Accessed 10 May 2018
22. Encyclopedia Britannica (nd) Jewelry—the history of jewelry design. Encyclopedia Britannica. https://www.britannica.com/art/jewelry/The-history-of-jewelry-design. Accessed 30 Apr 2018
23. Epoch Times (2013) How jewelry production hurts the environment, eco-friendly options. Epoch Times. https://www.theepochtimes.com/how-jewelry-production-hurts-the-environment-eco-friendly-options_415861.html. Accessed 7 May 2018
24. Fairmined (nd) Fairmined organization. Fairmined. http://www.fairmined.org/. Accessed 7 May 2018
25. Fashion Network (2018) Jewelry companies team up with IBM on blockchain platform. Fashion Network. http://uk.fashionnetwork.com/news/Jewelry-companies-team-up-with-IBM-on-blockchain-platform,972282.html#.WxEP3UgmA2z. Accessed 28 May 2018
26. Fisher T (2004) What we touch touches us: materials, affects and affordance. MIT Press Journals. https://www.mitpressjournals.org/toc/desi/20/2. Accessed 23 Jun 2018
27. Fletcher K, Grose L (2012) Fashion and sustainability—design for change. Laurence King Publishing, London
28. Floyd F (2018) The feel of fabrics, design for longevity. Design for longevity. https://designforlongevity.com/articles/the-feel-of-fabrics. Accessed 2 Jul 2018
29. Future Learn (2018) Fashion and sustainability course. Future Learn. https://www.futurelearn.com/courses/fashion-and-sustainability/1/steps/301108. Accessed 14 May 2018
30. Golani A, Wygnańska Z (2014) Beyond ornamentation. Jewelry as an aspect of material culture in the ancient near east. University of Warsaw, Warsaw
31. Goldenberg S (2014) Naomi Klein: 'We tried it your way and we don't have another decade to waste'. The Guardian. https://www.theguardian.com/books/2014/sep/14/naomi-klein-interview-capitalism-vs-the-climate. Accessed 4 Jul 2018
32. Gorjanc T (2018) New bioethics of luxury—how will advances in tissue engineering technology redefine today's concept of luxury? In: Bruggeman D, King I, Pildner K, Smelik A, Teunissen J, Eijnde J (eds) Fashion colloquium: searching for the new luxury. ArtEZ University, Arnhem, pp 58–59
33. Guardian (2013) The challenges of global gold across the supply chain—in pictures. The Guardian. https://www.theguardian.com/sustainable-business/gallery/challenges-global-gold-supply-chain-pictures. Accessed 30 Apr 2018
34. Hackspiel-Mickosch E (2016) Festivities at court: places of vestimentary luxury and royal performance. T and F online. https://www.tandfonline.com/eprint/jeb6TCIJDI9hErCSREPr/full. Accessed 10 May 2018
35. Harper D (1987) Working knowledge: skills and community in a small shop. University of Chicago Press, Chicago
36. Harper K (2018) Aesthetic sustainability. In: Bruggeman D, King I, Pildner K, Smelik A, Teunissen J, Eijnde J (eds) Fashion colloquium: searching for the new luxury. ArtEZ University, Arnhem, p 13
37. Helfand J (2016) Design: the invention of desire. Yale University Press, New Haven

38. Hepler L (2015) Tiffany & Co.'s new CSO on cleaning up the jewelry supply chain. Greenbiz. https://www.greenbiz.com/article/tiffany-cos-new-cso-polishing-jewelry-supply-chain. Accessed 5 Jul 2018
39. Hicklin M (2016) Human rights in supply. Chater for compassion. https://charterforcompassion.org/human-rights-in-supply-chains-human-rights-watch-cci. Accessed 29 Apr 2018
40. Human Rights Watch (2018) Jewelers should come clean. Human Rights Watch. https://www.hrw.org/behindthebling. Accessed 6 May 2018
41. Human Rights Watch (2018) The hidden cost of jewelry I Human rights in supply chains and the responsibility of jewelry companies. Human rghts watch. https://www.hrw.org/report/2018/02/08/hidden-cost-jewelry/human-rights-supply-chains-and-responsibility-jewelry. Accessed 6 Jun 2018
42. Kimberley Process (2017) Final Communiqué Brisbane, Australia. Kimberley Process. https://www.kimberleyprocess.com/en/2017-final-communiqué-brisbane-australia. Accessed 10 May 2018
43. Leach A (2018) Copenhagen fashion summit 2018 recap. Highsnobiety. https://www.highsnobiety.com/p/copenhagen-fashion-summit/. Accessed 2 Jul 2018
44. Malinowski B (1948) Magic, science, and religion. The Free Press, Glencoe, Illinois
45. McNeil P, Riello G (2016) Luxury a rich history. Oxford University Press, Oxford
46. Miller D (2010) Stuff, 1st edn. Polity Press, Cambridge
47. Miller D (2012a) Material cultures. Taylor and Francis Publication, Hoboken
48. Miller D (2012b) Consumption and its consequences. Polity Publication, Cambridge
49. Miller EL (2017) Luxury in Europe 1600–1815: negotiating narratives 2010–15. Researchgate. https://www.researchgate.net/publication/321222543_Luxury_in_Europe_1600-1815_Negotiating_Narratives_2010-15. Accessed 28 May 2018
50. New S (2017) The transparent supply chain. Harward Business Review. https://hbr.org/2010/10/the-transparent-supply-chain?autocomplete=true. Accessed 10 May 2018
51. Niemtzow E (2018) Disrupting luxury: creating resilient businesses in times of rapid change. B.S.R. https://www.bsr.org/en/our-insights/report-view/disrupting-luxury-creating-resilient-businesses-in-times-of-rapid-change. Accessed 8 Aug 2018
52. Nuwer R (2014) Future—what is the world's scarcest material? BBC. http://www.bbc.com/future/story/20140314-the-worlds-scarcest-material. Accessed 11 May 2018
53. Ogilvy D (1985) Ogilvy on advertising. Vintage Books, New York, pp 170–243
54. Pechstein A (2018) Biomimicry: what nature can teach us about circularity, value and innovation. In: Internationales Design Zentrum Berlin (eds) Circular design forum. Berlin
55. Phillips C (2008) Jewels & jewellery. V&A Publishing, London
56. Pinkhasov M, Nair R (2014) Real luxury: how luxury brands can create value for the long term. Palgrave Macmillan, Hampshire
57. Porter M (1985) Competitive advantage creating and sustaining superior performance. The Free Press, New York
58. Responsible Mining Index (2018) Index. Responsible Mining Index. https://responsibleminingindex.org/. Accessed 10 May 2018
59. RJC (2012) Case-studies. RJC. https://www.responsiblejewellery.com/members/case-studies/. Accessed 4 Jul 2018
60. RJC (2020) Lead with purpose in challenging times—progress report May 2020. RJC. https://www.responsiblejewellery.com/wp-content/uploads/RJC_Progress-Report_May-2020.pdf. Accessed 16 June 2020
61. Rudnicka E, Mamros L, DeRiggi B, Munshower B (2011) The diamond supply chain. Paper presented at the 8th Latin American and Caribbean conference for engineering and technology, Arequipa, Perú, 1–4 June 2010
62. Sawyer J (2018) ASKET's labeling standard discloses every step of manufacturing. Highsnobiety. https://www.highsnobiety.com/p/asket-labels-explain-manufacturing-process/. Accessed 2 Jul 2018
63. Sennett R (2008) The craftsman. Penguin books, London
64. Thomas D (2007) Deluxe: how luxury lost its luster. Penguin Press, New York

65. Verde D, Hanton K (2011) Positive luxury, the butterfly mark. Positive luxury. https://www.positiveluxury.com/about/. Accessed 10 May 2018
66. Verde D, Hanton K (2018) The world is ruled by emotions: Breaking tradition the millennial way. Positive Luxury 2018 Predictions Report. Positive luxury. https://static1.squarespace.com/static/59d752618dd0418b735bac93/t/5a65e742e2c4834cb74da27d/1516627791739/2018_Predictions_Report_Positive_Luxury_OwGpCBm.pdf. Accessed 24 Jun 2018
67. Voyer BG, Beckham D (2014) Can sustainability be luxurious? A mixed-method investigation of implicit and explicit attitudes towards sustainable luxury consumption. The Association for Consumer Research. http://www.acrwebsite.org/volumes/1017922/volumes/v42/NA-42. Accessed 31 April 2018
68. Welsh A, Chadprajong-Smith J (2018) Fashioning tradition—the Tai Lue Project. In: Bruggeman D, King I, Pildner K, Smelik A, Teunissen J, Eijnde J (eds) Fashion colloquium: searching for the New Luxury. ArtEZ University, Arnhem, pp 72–73
69. WGSN (2018) Sustainable strategies—jewelry. WGSN Publications, London

Tanzanites: The Maasaï Sustainable Dilemma for the Rarest Gemstones

Florent Vincent, Ivan Coste-Manière, and Marc Basseporte

Abstract Tanzanite, a unique gemstone that can only be found in Tanzania under difficult and rough conditions, is actually increasing in fame due to its scarcity and its deep vivid blue color. However, for every newly founded gemstone, concerns are arising toward the sustainability of its supply chain process. Discovered in 1967 by a Maasaï tribesman, the tanzanite gemstone faced several challenging times and thus making unstable its presence on the jewelry market in the last 20 years. But actually, what do we know about tanzanite except that it's coming from Tanzania? This study aims to examine the global climate of a tanzanite mining environment, understand areas that lack development and see how Tanzania is strategically repositioning tanzanite into the economy of the country in a sustainable way. The objective is to demonstrate that Tanzania is a resourceful country in terms of culture and history. This African nation should be known as a country that is able to manage efficiently and respectfully the process of mining and trading tanzanite to enforce its local development. Starting from the global context of tanzanite mining, its history, and background, this study keeps on analyzing the structure of stone distribution. It

F. Vincent · I. Coste-Manière (✉)
SKEMA Business School, Sophia-Antipolis, Valbonne, France
e-mail: ivan.costemaniere@skema.edu

M. Basseporte
Arusha, Tanzania

I. Coste-Maniere and M. Á. Gardetti (eds.), *Sustainable Luxury and Jewelry*, Environmental Footprints and Eco-design of Products and Processes,
https://doi.org/10.1007/978-981-16-2454-4_7

furthermore showcases the different improvements that have been made to increase the safety of miners as much as the transparency of the tanzanite market and provide work opportunities for Tanzanian inhabitants. Finally, by providing a global understanding of the current situation of the tanzanite business, this research presents foresight to improve the future of tanzanite trade and establish efficient strategies to secure the market. The results of this research highlight the strong importance of culture and local communities in the tanzanite mining industry which created disputes between small-scale and large-scale mining industries. From the beginning of tanzanite mining, multinationals and foreign countries were leading the Tanzanian market of gemstones. Tanzania's future horizon is now building on partnership and fair deals to clarify the tanzanite trade, thus reducing violence. The challenge was to redirect the wealth from abroad into Tanzania broader. To this fact, the Tanzanian government puts in place taxation programs for unpolished stones to enhance the local development of cutting skills. In partnership with associations, NGOs, and the main mining companies, they provide educational support for locals to develop the required skills and to allow people accessing properly into the tanzanite business. The results show that the tanzanite mining environment changed significantly since its discovery. Future perspectives for tanzanite are arising due to its growing interest in jewelry from other countries. The development of new technologies such as blockchain might help to secure miners and make tanzanite pathways more transparent. The respect of traditions and cultural heritages of Tanzania is very important to improve the stability of the market and benefit every stakeholder. Following these assumptions, the Tanzanian economy has a bright future ahead if the government will still invest into the management and control of the tanzanite business. Taking Botswana as a reference for their economic growth boosted by the diamonds industry, Tanzania could apply the same approach with the unique and fascinating tanzanite.

Keyword Tanzanite · Sustainability · Small-scale mining · Tanzania distribution · Traceability · Gemstone · Heritage

1 Introduction

Nowadays, the beauty of tanzanite gemstone is shining through the US market and starts quietly to be known within the European market. Its African cachet, its extreme rarity about 100,000 times rarer than diamonds, and, most of all, the strong power of its dichroic blue hue makes tanzanite stones very valuable. However, where valuable stones gain in popularity, conflicts and controversy might emerge quickly.

This paper will analyze the different steps of the supply chain that will allow the tanzanite to become more sustainable within the Tanzanian market itself. Establishing the current situation on tanzanite mining in Tanzania will help understand the underlying polemic. Hence, it will be possible to determine the following: What are the available solutions to overcome the obstacles that may crop up. How to benefit from ethical mining processes to help the local community and develop a long-term vision. How current technologies, project, and habits could enhance the efficiency of local miners. Finally, we will consider this fairly new gemstone could impose itself on the markets and what could change within the tanzanite market taking into consideration that tanzanite is part of the history and heritage of the local Maasaï community.

The three stages will therefore be analyzed in more detail below. They will raise the various concerns that tanzanite industrial and small-scale mining have generated and how local communities reacted to all these changes. A better understanding of the tanzanite supply chain will enable us to assess what kind of actions could be implemented to establish tanzanite trade and traceability more ethically.

In the first part, this research will focus on mining conditions, tools, and techniques already in place. It will highlight the consequences of the massive arrival of small-scale miners who wish to make a fortune in tanzanite, and the contrast with highly developed commercial mining facilities. Finally, we will focus on the practices and initiatives of the government and private companies developed to help support the tough working condition of mining.

In the second part, this study will focus on several methods that promote a more ethical distribution of the tanzanite trade. It will start from the source identification of uncertainties that hang over both national and international tanzanite trade. Next, the research will explain how education can provide solutions that might improve sales and distribution environments. The growing interest in the authentication and traceability of stones will enhance the transparency within the supply chain. Community projects will also help local communities through education with tanzanite business coaching, empowerment for Maasaï women, self-development within society, and most importantly by improving the quality of life through the creation of hospitals, orphanages, and schools.

Finally, tanzanite is no longer used only for mining and we see a diversification of activities linked to the stones. Auction centers, museums, events, and flagship stores appear to overall increase recognition as well as buyer's awareness. Thus, this part of the research will provide insights to be developed and boundaries to set to ensure sustainable and ethical exploitation of these gemstones in the long run in an ever-expanding market.

In the third and last part, this research will try to associate the strong Maasaï history and culture with the present challenges of the tanzanite trade, and how social ties have evolved following the rise of the mining operations. Generational conflicts might arise depending on how locals consider the flourishing tanzanite business compared to the human cost it entails. Finally, this research will provide a global vision of what are the future perspective in terms of tanzanite sustainable development and its opening to the European and Asian markets (Fig. 1).

Fig. 1 COSTE-MANIERE I (2017) Fashioned Tanzanite displayed as loosed stones

2 Part 1: Mining Operation and Controversy

2.1 General Information

Almost unknown until 1967 (but by Tiffany and some customers in the US market), tanzanite was discovered in northern Tanzania after a ravaged savannah. The vanadium-rich stones were sufficiently heated by the fire to turn bright with an exceptional violet–blue hue. A local Maasaï tribesman found a piece of this new specimen near Mount Kilimanjaro and brought it directly to a rich gold prospector who was looking for rubies at that time. After being alerted, Manuel d'Souza immediately registered for mining claims.

Even though tanzanite was discovered recently compared to today's most popular gemstones, this violet-to-blue mineral appeared 585 million years ago near the foothill of Mount Kilimanjaro, resulting in a junction between the tectonic plate activity and fierce heat in that area. Nowadays, the gemstone is known to be found only in Tanzania in a ridiculously small mining zone near the Mirerani hills of the Manyara region.

Tanzanite is a relatively new-colored gemstone. It comes from a variety of mineral zoisite which belongs to the crystal family. The main specificity of tanzanite lies in its pleochroism. The gem has different colors depending on the crystallographic direction from which it is observed. Tanzanite gemstone is said to be trichroic, which means that the same crystal shows three different colors from several directions. From one direction, it can be blue, purple to reddish from another view and bronze from the last direction. The well-known blue color of commercial tanzanite is usually enhanced with heat treatments in laboratories. The gemstone is made of a small amount of vanadium, which is a blue-silver-grey metal often used as a steel additive. Lab-heating the zoisite gem results in altering the oxidation level of vanadium which produces the desired blue color by dissolving its brownish tone. However, artificial heating has a significant devaluation effect on the gemstone and hence on its price. Therefore, a natural medium dark blue and untreated tanzanite is high-valued and sought-after by jewelers although it is quite rare.

The stone was not always been called "Tanzanite". This name was given by Tiffany & Co who introduced it to the market in October 1968. It was inspired by the origin country of this ancient "blue-violet zoisite". Indeed, Tiffany & Co recognized at this time the potential of tanzanite and built up a strong promotional campaign. The American jewelry company then became the main distributor of tanzanite and contributed to its rise in popularity across the world. Today, the gemstone is ranked second within the most beloved blue gemstone list after sapphire. In 2002, tanzanite reached another level of notoriety while the American Gem Trade Association selected it as one of the December Birthstone. Considering that the official birthstone list had not been updated since 1912, this nomination was a big step for tanzanite, which now possesses a strong position among the gemstone universe rivaling the Big 3 (emerald, ruby, and sapphire).

Many Tanzanians have worked in the mining industry, particularly tanzanite mining. Due to its growing fame and the potential income earnings that it can bring, this business has attracted many people who were ready to risk it all in gem mining. Obviously, when a coveted resource comes into being, tensions arise leading to crowded rallies around the mines. These people seek to generate some income to support financially their families.

2.2 *Nationalization of Tanzanite Mines*

Mines were nationalized in 1971. Nonetheless, the period between that time and 1990 is known for a downfall in production because of irregular mining with theft or even informal artisanal activity. External companies are establishing themselves on the Tanzanian market by associating a joint venture with local state-owned companies as was the case for STAMICO and TanzaniteOne for which the government and the private company each hold 50% since December 2013. The mines are entirely owned by the state which derives royalties from them. Regulations are difficult to put into practice and remain limited due to an unstable and fraudulent environment. The arrival of international private mining groups with their technologies and know-how to deal with local miners established for generations increased tensions. Competition is raging between private mining companies like Kilimanjaro Mines Ltd or TanzaniteOne (a joint venture of STAMICO) and small-scale miners.

The tanzanite exploitation in the shadow of Mount Kilimanjaro is divided into four blocks to separate companies' owners and allocate properly the zone of exploration. The latest figures state that 70,000 people are involved in blocks A and C dedicated to a private company and blocks B and D for small-scale mining. Boundaries of mining areas have helped to mitigate the pressure present on-site due to a very unstable climate from numerous smuggles and robberies. The tanzanite mining area is not large, around 14 square kilometers. Sometimes, within the underground maze, it happens that small-scale miner tunnels meet the exploitation tunnels of private companies during exploration. This happens quite often due to a lack of technology to indicate properly the limit of the prospection areas. Blocks being stuck together; it makes this kind of incident happen regularly in the region and it led to chaos where settling of scores and explosion occurred.

2.3 *Tensions and Insecure Mining Environment*

In order to ensure the safety of minors and limited robberies within the A and C, the Tanzanian government and TanzaniteOne decided to build a wall, at the end of 2017, separating the blocks in order to facilitate access control and leaving the mines. Indeed, pressure is high as the article written by Antony Zagoritis quotes: "*This led to conflict between the small scale Tanzanian miners who had traditionally mined the area and the foreign newcomers who they perceived as exploiting their resources and repatriating profits with little direct benefit to the Merelani community.*"[1] This statement really summarized the situation around Merelani, and we can understand how deeply they feel breached.

Undoubtedly, small-scale mining practices are well known all around the world and it is a common subject of dispute when there are valuable resources to exploit. For example, in Sri Lanka, the government decided to retain access to sapphire mines only for small-scale miners to enhance the on-site labor force.

Women and young miners are also part of the small-scale mining community and have a strong impact on the management of the mining process. According to the book "Between the Plough and the Pick: Informal, Artisanal and Small-Scale Mining …" published by Kuntala Lahiri-Dutt, the amount of population involved in small-scale mining is approximately 550,000 persons: "*Eftimie* et al. *(2012) reported that there were 550,000 small-scale miners in Tanzania, of whom 25 per cent were women.*"[2]

Adding to this, "*A 2003 South African report placed the number of child labourers in Mererani at 22,500 (Kinabo 2003).*"[3] Following the research of Groves, two-thirds of the child laborers were under the legal age of 16, working in the mine, and a minority of them had never been to school.

These figures are quite obsolete now, but they help in understanding the major role played by the women and the young labor force. Children and teenagers even had dedicated tasks because they were smaller and more agile. They were used in special cases to find difficult-to-access crystals in small drifts; hence had their nicknames "snake boys" or "nyokas" in Tanzanian dialect. They might have also helped to administer food and water rations to the miners and to monitor that the miners were not stealing the freshly found gems.

Technological improvements allow the wealthiest mining businesses to acquire X-ray scanners which, once installed at the entrance and exit of extraction areas, ensure that no stones are stolen. However, we will observe that the technological and substantive gaps necessary to establish optimal safety in the mines depend whether on the prospection area dedicated to small-scale or private miners. For a multinational like TanzaniteOne, the mining conditions seem overall rather secure. They use winches with three anchoring brakes to avoid any accident during the lifting of the ore and transport of the food. High-tech sorting machines are used to identify gems among the excavated ore. The production in larger quantities allows greater returns on investments but it receives break and even pressures to remain profitable economically. More staff devoted to security checks, machine control, and maintenance ensures significant protection from harm.

On the other hand, small-scale miners face harsher mining conditions and stressful working environments. Access to sufficient quality of air to work in good conditions is one of the main problems of the deep depth mining industry. Indeed, some drifts

---

[1] [1].

[2] [2].

[3] [2].

and veins take miners up to 1 km underground. Air evacuation is often complicated and tones of the ore that are extracted spread toxic fogs. Blasting explosives leave in the air dust, chemical, and rock pieces which are dangerous for the breathing of workers due to non-appropriate ventilation technologies. Knowing that some miners can work up to 18 h a day, ventilation of underground areas is, therefore, a major problem and affects the respiratory capacity of miners.

The lack of security and poor working environment are not in favor of small-scale miners due to the huge gap in financial and technical means between private and small-scale organizations. The tension is not helping to abolish violence onsite: "*Several miners and a TanzaniteOne employee have died in these disputes. Flash-flooding of mines and faulty air pumps have also led to deaths in Mererani. Tanzanite has thus become both a commodity fiction and a commodity nightmare.*"[4]

Drugs and prostitution—among others—also arise when poor conditions of labor appear. The locals, therefore, do their best to work in these tough conditions aiming to increase their average revenue. Finding gems is not only the main challenge of small-scale miners, as once the gem is extracted they will need to find buyers. Indeed, the presence of numerous individual miners means that there are many stones available on the market that need to be appraised and sold.

On top of that, HIV takes profit over the high density of people and the unbalanced number of men and women in the Mererani area. "*In 2005, the prime minister at the time, Edward Lowassa, said Mererani had a HIV prevalence rate of 16.4 per cent.*"[5]

The rise of illegal mining practices and the high labor force capacity of Tanzanian small-scale miners have saturated the market with many mid-range quality tanzanites. "*Many workers sell their minerals at lower than market prices to middlemen, some of whom sponsor their operations. The incomes of such miners are usually below the poverty line, further reinforcing their poverty cycle.*"[6]

2.4 Infrastructure and Legislation to the Benefit of Local Communities

Small-scale miners are the source of the evolution of tanzanite stones in the Tanzanian market. They fought for their right to dig on their territory, and now they need to fight to keep their jobs and preserve that precious monopole and uniqueness of the tanzanite place of birth.

As a result, several initiatives have been put in place by the government and private mining groups to maintain the economic momentum generated by the trade in the tanzanite mining. More generally, these actions are aimed at improving the security of minors and trading conditions. They established medical centers to ensure and control the health of minors "*Tanzanite One teams up with some doctors from*

[4] [2].

[5] [2].

[6] [3]

Kilimanjaro Christian Medical Center- KCMC to set up an annual Eye Camp for the miners."[7]

Private companies, which must remain profitable, have the need to manage their units of miners in operation. The efficient management of the number of miners required for prospecting will be adapted to market demand and the political situation. "*January 2016, TanzaniteOne, now controlled by Sky Associates of Arusha, laid off half of its workers. Citing reduced production of tanzanite (50 per cent during fall 2015).*"[8]

Fostering artisanal skills is the main objective to save jobs and create value with Tanzanian know-how and heritage of gemstones. Having the exclusivity from the outside being the only place on earth where tanzanite is drilled, the Tanzanian market needs to build on its inclusivity. It not only differentiates itself from the competition but also builds a reliant and sustainable chain of production that can positively impact the economy of the country. In addition, concentrate on the local market by reducing mining hostility and establish fair-trade and quality of the working environment.

Several laws implemented in the last 5–10 years follow this objective. In July 2010, a law was imposed against the exportation of rough tanzanite that weighs more than 1 carat. This promotes the development of local gem identification skills and cutting and faceting skills before inserting worthy gems on the market. "*The Government of Tanzania has also been promoting value addition of mineral products and this is done by reducing the tax. If a miner adds value to its products, the royalty is only 1%, otherwise it is 5%.*"[9] This tax encourages the cutting and polishing facilities directly on Tanzanian soil and avoids the exploitation of rough stones from major specialized foreign entities situated in India or Thailand for corundum (sapphire and ruby). "Actions such as tax cuts encourage the registration of tanzanite miners."[10]

3 Part 2: Ethical Distribution

Tanzanite mining has generated many quality gems reaching exceptional values in US dollars. However, it is easy to observe that there is a certain lack of control in the management of value and financial flows. There is a lack of information transparency because valuable gemstones can generate high profits and black-market revenue. This second part will therefore focus on the possibility of implementing a more ethical distribution in the business of tanzanite. This research will revolve around the organization of Tanzania which is trying to keep wealth, value creation, and skilled workers within the country. It will present the main stakeholders in the tanzanite business; how the lack of transparency in the exchange and traffic of gems can be improved, thanks to the help of new technologies; and how education is causally

[7] [4].
[8] [2].
[9] [5].
[10] [6].

linked to a more ethical future in the distribution of tanzanite. There is a need to resolve general conflicts of interest in the tanzanite industry. Finally, we will review the importance and role of each citizen to act more responsibly within the tanzanite business trade; and the role of education, financial, and physical aids to change the vision of the tanzanite market into a sustainable and qualified mining area in which clients have faith and trust in it.

3.1 Workers—Stones Dealers—Stones Extractors

All the actors in the distribution channel have their own roles: stone dealers, stone extractors, and workers. We will try to understand the relationship between them and how it can illustrate the lack of transparency in gemstones trades.

The most important stone dealer is certainly TanzaniteOne Mining Ltd which extracts tanzanite only. Currently, there are far fewer stone dealers and stone diffusers for tanzanite than any other market for colored stones. This could be entitled as a monopoly market considering the fact that Tanzania has the only exploitable tanzanite deposit in the world, which significantly limits the number of dealers since the mining area is quite limited in size.

The government is contributing mainly to block C where the most important stone dealer TanzaniteOne is installed leaving a few resources for the proper development and establishment of secure working environments in other blocks where the majority of small-scale miners are extracting the ore.

What is interesting to see in this part is why two companies, Tiffany&Co, the famous American jewelry house, and TanzaniteOne, are sharing the monopoly. We will also discuss how this extremely rare stone is used in the majority by Tiffany & Co which is the main stone spreader of tanzanite in the world. It is understandable that tanzanite is exploited by Tiffany & Co since it is the company that brought out the stones in 1968 and used it for the first time in one of its creations from Donald Clafin's: A floral brooch with diamond, emeralds, and tanzanite that weighs around 47.85 carats.

If we make the comparison with the diamond and its rise, we can therefore ask the question of why the tanzanite, a stone 1000 times rarer than the diamond, did not follow the same world ascent. Part of the answer is in the stone dealer's monopoly, which for one reason or another the majority only wants to do business with Tiffany & Co. But something strange might arouse suspicions. Why does a private company such as Tiffany & Co not deal more with other stone diffusers like Chanel, Dior, Cartier, etc. to increase the awareness of tanzanite worldwide? Perhaps, it is an artistic will or maybe the monopoly is so obvious that the competition is crushed, which makes access to this stone difficult.

In an advertisement in 1970s, Tiffany & Co claimed to be the only stone spreader of tanzanite: "Tanzanite can now be found in significant quantities in only two places in the world. In Tanzania and Tiffany's."[11]

Coming back to TanzaniteOne, it was originally a company created by Richland Resources Limited. Luckily, Richland Resources Limited also owned a company called Richland Gemstones, whose aim was to sell jewelry mainly made of tanzanite which comes directly from the mines belonging to the company. Richland Gemstone was so far a stone diffuser that promoted a direct chain of tanzanite jewelry by promoting Tiffany & Co's jewelries until the website was shut down for an unknown reason. The mining monopoly seems to be then in the hands of the giant Richeland Resources, or as we must say "TanzaniteOne Mining LTD", which is even listed on the London Stock Exchange as a company. It shows its strong financial power and negotiation capabilities. Tiffany & Co, on the other hand, was responsible for the design conception and the marketability of tanzanite jewelry since 1969. To this fact, they designed an entire supply chain in which they can control from the mining area to the jewelry store the demand and stock of tanzanite. The lack of clarity in the traceability of stones makes this monopolistic market questionable.

3.2 Relocation of Tanzanite Value Chain

Wealth transfers from Tanzania to the rest of the world with tanzanite commerce are imperative issues that Tanzania should address. The value of tanzanite production per year is about hundreds of millions of US dollars. However, due to the numerous illegal gem traffic, the lack of transparency for the real value creation generated by small-scale mining created obvious accounting inconsistencies. In 2017, "*over 8 shilling trillions have been generated from tanzanite according to government. However the TRA (Tanzania revenue authority) reported 400 shilling billions which represent 5.2%.*"[12] The 5.2% reported represents around 172 million US dollars. What happened to the 94.8% remaining? Here are the major issues for all the diverse sectors of precious stones mining. Smuggling is deeply enrolled in the life of a colored stone, and this is why it becomes extremely difficult to have pure "green and ecofriendly" mined gemstones.

The fluctuation of value in gemstone trading is difficult to track, and trends showed that the vast majority of tanzanites are shipped directly to the US market. About 80% of tanzanite goes to the US market. Tanzania, therefore, needs to focus on the inclusiveness of its market. Tanzanite Experience, a non-profit organization created by TanzaniteOne, "*is on the road to a major repositioning of the local gemstone industry.*"[13] Support for the inner development of qualified gem trader and "fashioner" (terms used to describe a person in charge of cutting and fashioning rough

[11] [7].

[12] [8].

[13] [9].

into polished stones) are part of the responsibilities of mining companies. Centralized expansion can be cost-effective in the long term and may facilitate the social and sustainable dynamic of change.

3.3 Education

It is important to support a more direct chain of buying and training programs for individuals involved in the tanzanite small-scale mining. Joining the business of tanzanite mining might be sometimes done as a default of elder generation decision or to earn more money. But most of the small-scale miners have exceedingly small knowledge or education. They might know how and where to mine but the business part, where they need to sell their own gems, can be problematic for them.

Many small-scale miners changed their previous jobs for mining. Taking the example of "Prisca",[14] a Tanzanian businesswoman who chose to convert into a small-scale miner because her business was not making any profit anymore. Mining, even with tough conditions of work, long hours of labors, and the need to purchase mining equipment brought her a more regular return on effort invested than her previous jobs. This highlights very clearly how tanzanite mining is important in the region of Mererani at a point where small businesses need to close. Unfortunately, "Prisca" is facing many challenges about selling her stones such as low price per gem. Small-scale miners have more difficulty in negotiation. The first reason is that even the cheapest price per gems mined is always more revenue than their previous jobs. They are lacking expensive mining equipment and education. They do not have any bargaining power because they do not have any other solution rather accept the offer made to them. There are plenty of small-scale miners in the same situation. They want to sell their rough stones to bring some money back to their homes. However, the market for low-carat tanzanite is already well saturated, making it more difficult to negotiate prices for individual sellers.

Many associations are focusing on empowering women labor force and autonomy such as Pactworld WORTH (Women Organizing Resources Together), Maasai Woman's Project from TanzaniteExperience enterprise, or also Tanzania Women Miners Association (TAWOMA). These organizations are "*local initiative empowering Maasai tribeswomen from the mining area of Merelani with their own business venture.*"[15] Indeed, as we have seen in the first part of this research, women represent almost 20% of the small-scale miner labor force and most of them wish to succeed in business as men, but they lack business management knowledge. These associations tend to unite women together in groups, allowing them to exchange their own experiences and feedback about mining. These associations provide material and teaching lessons on different topics relevant to interpret the current situation in the mine and how to deal with it.

---

[14] [10].

[15] [9].

For example, "WORTH" initiative from Pactworld, "WORTH brings poor, primarily rural women together in groups of 25 for integrated, savings-led microfinance, literacy, numeracy, group banking, and microenterprise development, while strengthening social networks and empowering members to have a voice in their communities."[16] Pactworld is a non-profit international company which focuses its business on sustainable development of challenging area, providing encouraging structure for the small-scale miner to develop their understanding of tanzanite trading. Pactworld brings insights about how to mine safely, how to keep their gems, and resell them at a fair price during market day. In association with GIA (Gemological Institute of America), they launched their action called "Empowering Artisanal Gemstone Miners in Tanzania" to share content and explain the real value of the tanzanite. They provided a guide to teach the fundamentals of gemology and gem identification as well as on-site expertise. "This helps miners better understand the quality and value of their products, which improves their position in the supply chain and their economic development opportunities."[17] Many projects are rising toward the shared goal of facilitating the implementation of small-scale miner to understand what tanzanite mining might provide for their future.

Education is a primary need that has to be filled in order to understand how the tanzanite chain of process is working. Indeed, we are talking about teaching small-scale miners on how to deal and make wealth from their mined stones despite all the parties involved in the supply chain. They must cultivate and exercise a new way of working in order to flourish over the long term. Educate at every stage of the process and participate in the establishment of an ethical line of distribution. No one is smart enough to understand perfectly how all the stages of tanzanite trading should be framed and how it should correlate altogether.

Thus, educational program should be established in relation to leaders of local community, leaders of international companies, and government to create didactic content relevant and adapted to the needs of each actor of the distribution chain. The educational program for small-scale miners should enhance mining safety, ethical behavioral responsibilities, and encourage the development of local talents providing development opportunities.

Sales representative trainings should show transparency and authenticity through the presentation of their products. Most of the time, gemstones followed a long and tough path from their mining to the design and final phase of the creation of a jewel. The salesman/woman should transmit his/her knowledge to the customer properly and explain what is involved when they buy tanzanite jewelry (Fig. 2).

Buyer's guidance is probably the third and most significant point to discuss following the demand of the current market. More people want to understand and clarify the way stones were processed and how relevant their purchase is. Choosing the real one approved with certification, with the best grade and the best price will vary following supply, demand, and steps in the supply chain where you buy the stones. If all these services could be covered during a sale presentation, clients will

---

[16] [11].

[17] [12].

Fig. 2 COSTE-MANIERE I (2017) Salesperson showcasing brand new-fashioned tanzanite

be able to buy quickly, share their experience of learning about the tanzanite story, and the loop of ethical distribution of stones is complete. Certainly, this strategy applies perfectly on paper but when it is implemented on the field it becomes another challenge to handle.

3.4 Certification and Traceability

Tanzanite is becoming popular in the colored stones family, thanks to its astonishing panel of three colors after treatments ranging from deep blue to royal blue and deep violet purplish. Its value is intensified, thanks to its beauty and rarity taking into consideration that there is only one place on earth where this stone can be mined, just behind the shadows of Mount Kilimanjaro. Its hardness of 6–7 on the Moh's scale makes tanzanite a perfect stone for all designers of jewelry houses. Since tanzanite has a weak toughness, it is therefore mainly fashioned on necklaces or bracelets and more rarely in a ring.

The tanzanite codes are consistent with those of Tiffany & Co: blue hue of tanzanite and cyan blue from tiffany. However, we observe that the Tiffany's monopoly around tanzanite is disappearing, leaving the place to new design creations coming from powerful jewelry houses staging tanzanite such as:

- **Chanel**. (2015) Chanel Set Cushion Cut Faceted Tanzanite, 18 K Gold, Tension Setting Ring
- **Dior**. (2018) « Dentelle Popeline » Bracelet. Tanzanite, 18 K Gold, Diamonds, Pink and violet Sapphires, Garnet, Tsavorites and Rubies.
- **BVLGARI**. (80s) Ear clips, Diamonds, Tanzanite
- **Chopard.** (2014) Temptations Collection, Ring. 18 K Pink Gold, 55ct oval-cut tanzanite, 11cts of Spessartine garnets, 3.6cts of Colored Sapphires, Pear-shaped sapphires
- **Buccelati**. (2014) Cocktail Ring Designed by Isabelle BULTEAU, 18 K White Gold, 15cts Tanzanite, brilliant-cut diamonds.

However, the diversification of stone diffusers requires carefulness about certification and traceability of stones that are becoming more and more necessary for firms to justify properly the transparency of their supply chain. Sourcing gemstones has always been a tremendous challenge for those companies. There is a multitude of stages of stone transition and exchange from the moment it is mined to the hands of the customers. Certification is definitely greatly beneficial to reduce smuggling, fake stone manufacturing, and counterfeits trade.

Certification is based on five criteria following the procedure of the Tanzanite International Grading Standard: weight, shape, cut, color, clarity, and edition number. It helps to identify where the gems come from and evaluate their value. Only expert gemologists can examine and grade the stones.

Tanzanite certification projects are repositioning from the inside of the country. It encourages building and growing local expertise before the stone travel worldwide. It also holds skillful Tanzanian gemologists back in Tanzania, develops job opportunities, and benefits from the increase of gemstones registration. Buying non-certified tanzanite is a dangerous bet to take. There could be even cheaper, but you will have no proof of origin, no information about the condition in which it was mined, and most importantly if it is really tanzanite, not a stone imitation.

Several companies stated that they are the pioneers in establishing certification processes for tanzanite. We can see that there are already several in action, for example: "*Tanzanite Experience is unique in that it is the only company in Tanzania with direct access to a certification lab.*"[18] Or even "*IMPACT PROJECT*" in Tanzania: "*We also provide analysis of certification, traceability, and due diligence as it applies to conflict-prone minerals in the Great Lakes region.*"[19]

Authenticating tanzanite based on these different physical characteristics is one thing. But to identify the source and the path taken by this same stone during the stages

[18] [9].

[19] 13.

of distribution is the main challenge in the mining sector. Nowadays, consumers are extremely sensitive to this kind of details since scandals have erupted because of the use of non-responsible stones (DeBeers and the "blood diamond" scandal). Jewelry buyers are particularly fond of knowing where the gemstone comes from and if it is not a stone from black markets. Traceability becomes crucial and "*a country-of-origin report is used in general to support a claim made about the geographical origin of a high-end gemstone (e.g. at auction).*"[20]

Traceability starts its journey from the increasing number of small-scale miners that record their mining process and come into the grid to sell legally their stones by gaining the same time visibility. Effective traceability is divided into four different parts and follows the model from the International Conference on the Great Lakes Region (ICGLR) in South Africa and Tanzania: "*1. Chain of custody tracking from mine site to export; 2. Regional mineral tracking using an ICGLR database; 3. Independent third-party audits; 4. An Independent Mineral Chain Auditor.*"[21]

Stages 3 and 4 ("Independent third-party audits; 4. An Independent Mineral Chain Auditor") represent internal and external due diligence required to determine traceability. Auditing the trade of gems is a way to check if the mining and trading procedure use fair and sustainable methods. It helps to determine if the environment where the stone was traded is legitimate.

Moreover, stage 2 ("Regional mineral tracking using an ICGLR database") manages the use of a database in order to retrieve the origin of the place where the stone was mined, under which name of company mining, when it was dug out, and by whom.

The blockchain approach, which is an information managing system, can visibly respond to this need. Mining blockchain initiatives are popping up considering the market demand on request of more transparency through traceability. It manages important amount of data linked to mining exercise. Blockchain can take into consideration various criteria to search and rewrite the story of the stones since it was mined. These projects are becoming more and more popular and could have a different use. Following a research from Laurent E. Cartier, Saleem H. Ali and Michael S. Krzemnicki "*today's laboratory reports offer origin opinions for certain coloured stones based on the scientific interpretation of their microscopic, spectroscopic and chemical properties compared to a reference collection of samples and the gemological literature.*"[22] Every investigation and results that a gemologist interpreted in previous research can be stored centrally in a database.

The following table presents some of the latest projects using blockchain management system to improve traceability at different stages in the gemstones industry.

[20] [14].

[21] [15].

[22] [14].

Initiative	Year founded	Target material	Chain of custody (or model provided)	Strategy	Claim made/aim of initiative	Supply chain segment
Tracr (De Beers)	2017	Diamonds	Blockchain traceability	Develop mine-to-finger blockchain for diamonds	Demonstrate traceability of diamonds from mine to finger via blockchain	From mine to end consumer via blockchain
Diamond Time-Lapse Protocol	2018	Diamonds	Permissioned private blockchain	Show the journey of a diamond to an end consumer via blockchain and app: option for manufacturers to track stock through manufacturing process via blockchain	Journey of a Diamond can be followed through manufacturing via blockchain and app	Manufacturer and retailer interface as well as consumer interface
Provenance Proof	2018	Colored stones	Blockchain traceability	Mine-to-finger blockchain for colored stones developed by Everledger and the Gübelin Gem Lab	Demonstrate traceability of colored stones from mine to finger via blockchain	From mine to end consumer via blockchain
TrustChain	2018	Gold and Diamonds	Permissioned Private blockchain	Offer traceability of diamonds jewelry via blockchain by working with selected certified miners, certifiers, manufacturers, and retailers	Provenance claims for the source of metals and diamonds used in jewelry items	From mine to end consumer via blockchain

Table extracted from: Laurent E. Cartier, Saleem H. Ali and Michael S. Krzemnicki. (2018) Blockchain, Chain of Custody and Trace Elements: An Overview of Tracking and Traceability Opportunities in the Gem Industry. The Journal of Gemmology.

Different approaches and methods of traceability could be implementable within the tanzanite business aiming to improve its ethical distribution. This second table coming from the same research showcases the different approach that blockchain traceability can bring, what could be their use, and how cost effective it is.

Traceability model	Approach	Level of traceability	Cost	General example	Gem example
Identity Preservation or Track-and-Trace	Certified materials and products are physically separated from non-certified materials and products at each stage along the supply chain	Highest	Very costly	Consumer would know exact farm from which a banana or salad was sourced	Exact mine of origin information is tracked through the supply chain
Bulk Commodity or Segregation	Separates certified from non-certified materials but allows mixing of certified materials from different sources. All producers must comply with the certification standards	High	Costly	An organic chocolate bar that contains cacao beans from various organically certified producers. Another example is Kimberley Process rough diamonds certified as "conflict free"	An aggregation of goods from one company that operates several mines: also useful for gem regions/countries and could be complemented by gemmological analysis
Mass Balance	Certified and non-certified materials can be mixed. However, the exact volume of certified material entering the supply chain must be controlled. Claims of 'this product contains X% of certified ingredients' can be made	Low	Slightly costly	If 20% of the total cocoa purchased comes from fair-trade sources, 20% of a company's chocolate bars made with that mix of cocoa can include the fair-trade certified label	Material from different mines (and certified and non-certified goods) can be mixed. Traceability information is lost

(continued)

(continued)

Traceability model	Approach	Level of traceability	Cost	General example	Gem example
Book and Claim	Allows all actors of a supply chain to trade in certificates for certified sustainable materials. Buying certificates allows retailers and manufacturers to claim that their business supports the production of sustainable materials. Claims of "this product supports the sustainable sourcing and production of essential commodities" can be made	Low	Reasonable	Companies wishing to make sustainability claims can purchase certificates (even though their goods may not be certified) that support sustainable production	A synthetic diamond manufacturer may buy credits and contribute to sustainable mining activities

Laurent E. Cartier, Saleem H. Ali and Michael S. Krzemnicki. (2018) Blockchain, Chain of Custody and Trace Elements: An Overview of Tracking and Traceability Opportunities in the Gem Industry. The Journal of Gemmology.

Companies committed to the gemstone industry nowadays are facing challenging times. They need to ensure a safe environment of work for their labor force and also provide a clear and detailed process of gem sourcing while training and increasing the awareness of their clients. Not only implement certification from the provenance of the stones but explain its signification and how to read it, helping, advising consumers, boosting confidence, reassuring the clients, explaining in a simple way the process of supply, etc. This is what will build up the responsibility of mining corporations and strengthen the gemstone buyer's trust.

Fig. 3 COSTE-MANIERE I (2017) Fashioned Tanzanite and rare fancy tanzanites displayed as loosed stones for semi-mounted jewelry

3.5 Undergoing Projects and Future Potential Initiatives

Finally, some resourceful projects are already implemented in Tanzania or could come out soon to develop a more ethical tanzanite distribution. Tanzanite Experience created a flagship store to promote tanzanite education and knowledge transmission purposes. "*The flagship store in the center of Arusha at Blue Plaza also has the* "*only tanzanite museum in the world.*"[23] The flagship store demonstrates the company values and perspective plans for the future of tanzanite, what they are standing for in terms of socials responsibilities facilitating the upholding of a close relationship with evolving clients' demand for new physical experience.

The market of second-hand luxury goods and loose stones trading are very promising for the future of tanzanite. The loose stone market could be launched within the second-hand store, development of sustainable and upcycled resources adapted to the client needs. However, the main issue still would remain to properly identify the stones path. It is indeed even more important because "*The advantage of buying loose stones is that you have more freedom of buying sets of stones that you can use to create sets of jewelry as you wish.*"[24] Provide the freedom of choice, guidance, and expertise on the combination of the potential stones combination and trusted certified mined tanzanite is the aim of an ethical chain of distribution (Fig. 3).

[23] [9].

[24] [16]

Auction is another alternative to maintain the valuation of tanzanite stones within the Tanzanian market. We can observe in the research "Between the Plough and the Pick: Informal, artisanal and small-scale mining in the contemporary world" auction plan started in 2008. "*In a hearing in Merelani in January 2008, representatives of the miners asked for in-country annual auctions to ensure that international gemstone buyers would leave their money in Tanzania and, presumably, not deal with sources in India or elsewhere (Juma 2008).*"[25]

Lastly, in the events sector, there is an established activity that enables to increase the awareness of gem trader and to understand the story and ambiguity that are still present in the tanzanite market and what are on the way to change. Tanzanite could profit from initiatives of other stones mining, which follow the same purposes of transparency and support. For instance, "*De Beers Group is piloting GemFair, a program that aims to connect artisanal and small-scale miners to the global market through digital technology and assurance of ethical working standards.*"[26] Gem fairs are definitely reliable activities to implement but actually due to political or even post COVID-19 crisis issues, it can be challenging to put in place which could move more toward a digital horizon.

4 Part 3: Apprehension About Tanzanite and Potential Future Sustainable Contingencies

4.1 Culture and Local Communities: The Roots of Tanzanite Mining

Discoveries of new gemstones have usually generated conflicts and pressures. Tanzanite gemstones are deeply involved in the local history of Maasaï residents, and the arrival of the international mining corporation aiming to excavate massively their precious tanzanite created tensions due to a gap in working and mining conditions (Fig. 4).

Poverty, constant disputes, and smuggling are issues that the government is trying to solve by settling a more responsible environment for the proper establishment and control of tanzanite trade. A limited area established for both private and small scale-miners and wall separation tend to improve slowly the security near the mines.

Moreover, Tanzanian miners are facing generational conflict leading to societal uncertainties toward the livelihood supply. Indeed, the new generation is questioning whether it is more profitable to go to work in the tanzanite mines in order to bring back money to their family as their parent did or they better go to school wishing to find other professional perspectives. These questions should not be raised, but the instability in which many local tribes live as well as the social traditions of their

[25] [2].

[26] [17].

Fig. 4 COSTE-MANIERE I (2017) Cultural Heritage Museum in Arusha

communities create this dilemma. Elders would prefer that their grandchildren follow the footsteps of ancestral traditions of searching for tanzanite, whereas parents will prefer to send their children to school so that they have access to an appropriate education. Thus, parents will go to the mines to acquire enough money and spare the harsh working conditions for their children.

In addition, working in the tanzanite industry can bring influential power in the region. "Elders strengthen their societal-level influence by endorsing young men's ventures to Mererani and connect young men's work at Mererani with community development and societal level benefits. This strengthens their influential authority in Maasai society, as well as their collective responsibility of securing common resources."[27]

Maasaï and locals are aware that the tanzanite trade brings them higher revenue in the short term, but it does not benefit them in the long term. The majority of high-quality mined tanzanites are sent directly abroad to be cut, faceted, and mounted on jewelry in notorious design houses.

Local communities mistrust the future of the tanzanite trade and expect real changes from the inside of Tanzania. Tanzanite mining companies build roads, hospitals, and orphanages but Maasaï's worry is what will happen in 30 years when the estimated capacity of tanzanite production will be terminated and that every external support will leave to another market of valuable gemstones.

[27] [18].

On the other hand, there are many inconsistencies in tanzanite trade wealth flows. To counter these dynamics, it becomes critical to centralize the tanzanite supply chain within Tanzania. Install a more direct and sustainable chain of buying by promoting education within every stage of the trade (miner, buyer, and seller). Build a responsible way of sourcing. Certifications will be allowed to record more tanzanite stones and miners, enhancing transparency in the tanzanite trade. Traceability and blockchain system will enable to follow the path of the stones and better manage its value chain. Sustainable sourcing can be therefore applied following a simple logistical process, taking into consideration the integrity of the workers, environment-friendly mining method, and waste management using prevention upcycled/recycled materials. Improve health and educational infrastructure in the region of Mererani to take care of citizens. Fortunately, organize an inclusive market that becomes essential; develop Tanzanian artisanal skills and environmental sensibility through diverse activities such as woman self-development associations; train on different growth drivers to improve their financial situation, their living conditions, and being able to develop themselves in this very challenging business of tanzanite.

However, to make significant changes, it requires government intervention. Accommodating ethical individual behaviors and generating and diversifying job propositions will increase livelihoods and enhance local consumption needs. The tanzanite flagship store, second-hand market, auction, and gems fairs are opportunities to exploit in order to increase the awareness of the tanzanite market and benefit the local economy. Creating companies to supervise and manage these transactions can make Tanzania not only the initial point of mining tanzanite but become a development and innovation hub for South African-colored stones.

4.2 Tanzania Facing the Growing Demand of the Chinese Market: Who Is Ruling the Game?

The symbol of tanzanite, called the Birthstone of December, also fits with Chinese values, as the gem is generally offered to a newborn child as "a family treasure", as a heritage for future generations regarding the scarcity of the gem. The strategic marketing campaign "Be Born To Tanzanite" (registered trademark of the Tanzanite foundation) is an example that is associated to the wish of The Tanzanite foundation to focus on a new market for the tanzanite trade.

Networking with prominent and progressive Chinese companies was therefore the more profitable evolution for Tanzanian traders. In September 2013, Chow Tai Seng Co Ltd and Richland Resources, a Tanzanian miner specialized in the mine-to-market supply channel for precious gemstones, signed an agreement to sell the finished tanzanite stones into China. From the mines, the rough stone is sold at buying offices in Arusha. Then, it travels to cutting facilities, almost exclusively located in India and the finished gems are then sold to Chinese jewelers.

Through this supply chain strategy, we can understand how important Asia in the tanzanite trade is. Nevertheless, as the retail jewelry network in emerging markets such as China is not yet well-established, a new form of networking emerged. Current relationships with sight holders appear more profitable to optimize interest in tanzanite. The main actor of the tanzanite trading between Tanzania and China remains, as stated previously, Tanzanite One Mining Ltd. But following its recent acquisition by the Hong Kong-based Sky Associates Ltd in 2015, we can wonder who is now ruling the game? This acquisition reveals the opaque evolution of the trade and position of Chinese actors.

The challenge is to convince the Chinese government to foster industrial development by subcontracting to local firms and enhancing education, and to develop technology transfers in the transport sector and the mining sector. The vice-president particularly asked for investments in the form of joint ventures and technology transfers to fight against soil looting and help Tanzania to process the ore before exporting it.

This new opportunity not only offered a stable economic growth for Tanzania but it also revealed new stakes, including dependence on Chinese demand, illegal trade and mining, as well as soil looting, leading to scarcity of resources. The Chinese demand is strongly influencing the market and the future of Tanzania's development, a double-edged evolution. In fact, after the 2010 Hong Kong Show—a trade show organized to promote tanzanite, an executive from Block C reported a general increase of 16.5% per carat and a top-grade increase of 29%. If the demand is currently constant, Tanzania could be affected by future lack of interest from Chinese consumers. Indeed, most of the luxury industry depends—between 30 and 40%—on the Chinese clientele nowadays. The establishment of further Sino Tanzanian exchange strategies should be controlled and develop certain perspectives that do not undervalue the growth of Tanzania by itself. Indeed, the overexploitation of the mines is progressively leading to scarcity of resources. This topic is an inherent issue, and it is for this reason Tanzanian economy should build other activities trying to become less dependent on the tanzanite market.

5 Conclusion

Tanzania has a tremendous luck to have this precious stone that is only exploitable on its soil. The goal in gemstone mining will remain in the capacity to contain the resources generated within the country. Companies involved in tanzanite mining should take many criteria into consideration. Reduce poverty, secure sensitive areas, and improve the living conditions of most Tanzanians that are still living in poverty. According the Poverty and Equality Databank[28] (2017), 49.1% of the population in Tanzania in 2011 was living under the poverty line. This data relies on a poverty headcount ratio at 1.90 a day. This tendency for high levels of poverty is likely

[28] [19].

to decline. In 2007, 55.6% of the population was living under this poverty line. In Tanzania, the poorest 20% of the population share 7% of the country's wealth, while the richest 20% share around 45% of the wealth. These inequalities are stabilizing for 15 years now.

Companies should establish a fair and balanced environment of mining areas between joint ventures and small-scale miners, thus decreasing the hostilities on-site. Respecting the cultural traditions and taking into consideration the strong generational influence of local communities is crucial.

Learning today's available latest technologies might help to establish a clear trade and safe mining environment for this new gemstone, strongly develop registration facilities, and raise awareness of tanzanite stones through tangible projects and digital presence.

Implement a future-oriented plan of action achievable in the short and medium term and establish a long-term vision.

Create infrastructure for the overall development of the Tanzanian economy that will not only remain reliant on the mining industry which created lots of wealth unfairly distributed. The fixed number of tanzanite will allow planning more strategically its mining future, and how to drive this business without draining the Tanzanian economy, but use it as leverage for growth.

Companies should answer to all these problems aspiring to an ethical distribution of Tanzanian jewels and being able to respect their involvement in corporate social responsibilities.

Joint ventures with the government allow a better comprehension of the real opportunities to consider. Corruption is unfortunately present wherever poverty is high, which is a major obstacle for the advancement of economic development and the prosperity of Tanzania. The government must track the tanzanite business from small independent producers to international traders, to make sure that the business will benefit the country. By taking serious decisions and implementing real regulations, the government could impact the tanzanite availability and price stability on its market.

The future of tanzanite will also be defined by the following economic trends likewise for every different stone: supply and demand. Changes in consumption and new concern about the product path required a distinct organization of the gemstone industry. The tanzanite market might face a decrease in the supply of tanzanite due to extremely limited quantities for its extraction, in parallel to a huge rise in demand, which impacts the price fluctuation. Because of its unique position in the world and only a few resources owned by international or independent miners, an increase in demand or a change in the offer will consequently impact the price. The interest from Western countries has always been remarkably high. In the USA, this demand is quite regular, being the biggest consumer of tanzanite. Its consumption represents 70% of the whole tanzanite stock, with an annual market of 300 million dollars.

In addition to a high demand coming from the West, new demand is emerging, which makes tanzanite resources even more desirable. Since the beginning of the tanzanite business, producers have been essentially dealing with Western countries. As we have seen, Tanzania and the USA are still quite close due to a long-term trading partnership with Tiffany & Co and Richeland Resources. However, new demanding markets, such as European, Caribbean markets, and more recently India, China, or Japan, have increased the potential demand for tanzanite due to its scarcity and exclusivity. Thus, today, the tanzanite position stands in the top 10 best-selling gems in the world, and this new influential demand will amplify the economic situation, driving prices to rise and make the supply to become increasingly limited.

Differences will be made on how the tanzanite will be marketed in a new market where it is not yet well known. The issues are based on the durability of the economic profitability and maybe some markets are already open to receive tanzanite, but the added value is not high enough to satisfy the industry owners.

Furthermore, the current demand is asking for ethical distribution and the use of new tech for sourcing and trace stones as well as high-quality stones which substantially makes the price rise and creates a decline in the availability of the finest stones. The most prized tanzanite (representing less than 1% of total production) will be fixed at premium prices; as a consequence, stones of lower quality will be set at the lower price. Because of the rise in demand beyond stone's availability and its impact on prices, potential investors are more and more fascinated by the gemstone. Investing in tanzanite is becoming a common practice as long as tanzanite stock is fixed, and tanzanite stone will probably gain value in the years to come. There is only one place in the world producing this gem for the moment and its production could end after 2050. The current price is between \$550 and \$700 per carat while probably going up to \$1,000 in a few years following the growing enthusiasm for tanzanite. More and more international brands (Chanel, Dior, Chopard…) are implementing tanzanite to their design collections. Tanzanite prices could escalate if the stone extraction becomes more and more arduous due to difficult access deep of tanzanite in the earth and other unexpected crisis such as the COVID-19 that the mining industry is currently dealing with.

The evolution of Tanzania will rely on government intervention and how Tanzania deals with the foreign direct investment that has a growing interest in Tanzanian resources. It is important to consider several possible solutions that could be taken to fight against resource extinction.

On the one hand, limiting the export abroad by offering production controls to only a few official operators could allow the limited stocks of resources to be preserved and centralized to one country: Tanzania.

On the other hand, assuring national and international controls with clear norms and standards for the whole trading system could make the market to be better regularized. For instance, whereas diamond is controlled by organizations such as DeBeers on the world market, there is less control or such a law system for tanzanite. In 2003, TanzaniteOne Mining Ltd, the leading miner, created The Tanzanite Foundation, a non-profit organization to promote the stone and take actions through the Tucson Tanzanite Protocol, aimed at implementing ethical and legal rules for the market to protect its limited resources (similar to the Kimberly Process for the diamond).

Finally, every luxury industry and mining operation has been deeply affected and slowed down during the COVID-19 crisis. Some donation projects from Gemology Institute of America (GIA) and TanzaniteOne are popping up to support small-scale

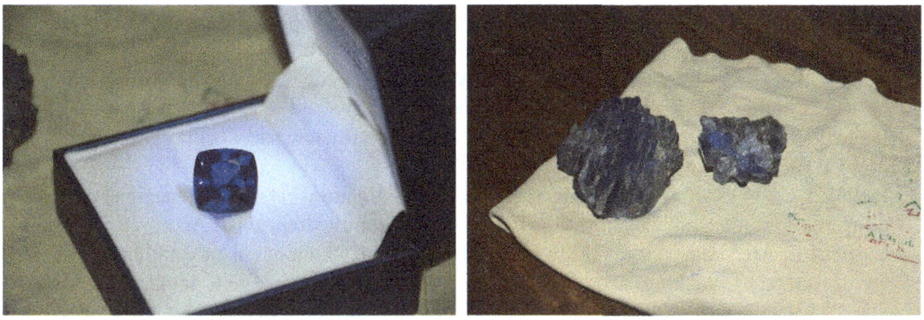

Fig. 5 COSTE-MANIERE I (2017) Stunning picture of rough and faceted tanzanite

miners but the future is still uncertain. Even if the current situation of tanzanite seems quite a bit complicated, there are still opportunities to explore. There is a lot to learn from the local tribes about tanzanite and their cultural background.

Fortunately, beautiful specimens are still to be discovered such as recently the two extraordinary tanzanites unearthed within the Saniniu Laizer's tanzanite mine in June 2020 that weighed 9.27 and 5.10 kg (Fig. 5).

References

1. Zagoritis A (2020) Tanzanite mining—large scale or small scale and the "Ethical" debate. In: GIA gemologist article. Available via The Rare Gemstone Company. https://www.therar egemstonecompany.com/gemology-articles/tanzanite-mining-small-scale-vs-large%20scale. Accessed 12 Jun 2020
2. Lahiri-Dutt K (2018) Between the plough and the pick: informal, artisanal and small-scale mining in the contemporary world. ANU Press. The Australian National University, pp 76–80. ISBN: 9781760461720
3. Reuben G, Zvidzai M (2015). Sustainable mining in Africa: standards as essential catalysts. ARSO Central Secretariat. Nairobi (KEN). http://www.arso-oran.org/wp-content/uploads/2014/09/Sustainable-Mining-in-Africa-Standards-as-Catalysts.pdf. Page 38. Accessed 07 Jul 2020
4. Medha P (2015) Tanzanite—"Gemstone of a Generation". Skara Jewels. https://skarajewels.wordpress.com/2015/12/11/tanzanite-gemstone-of-a-generation/. Accessed 02 Jul 2020
5. United Nations Economic Commission for Africa (2010) Tanzania | ASM. African Minerals Development Center. https://knowledge.uneca.org/ASM/Tanzania. Accessed 02 Jun 2020
6. Daghar M (2019) Efforts to curb tanzanite smuggling make Tanzania shine. ENACT Africa. ENACT. https://enactafrica.org/research/trend-reports/efforts-to-curb-tanzanite-smu ggling-make-tanzania-shine. Accessed 25 Jul 2020
7. Roberts C (2018) Tantalising tanzanite: the ultimate blue gem. Www.Thejewellery editor.Com. http://www.thejewelleryeditor.com/jewellery/article/50-years-tanzanite-ultimate-blue-gem/. Accessed 10 Aug 2020
8. Chidawali H, Kolumbia L, (2017) Simbachawene resigns as JPM receives two damning reports. The Citizen. https://www.thecitizen.co.tz/News/Simbachawene-resigns-as-JPM-receives-two-damning-reports/1840340-4087216-11n3ov4z/index.html. Accessed 04 Jul 2020
9. Tanzanite Experience (2020) Meet the team behind our shops. The Tanzanite experience. https://www.tanzaniteexperience.com/about-us/ Accessed 28 Jul 2020

10. Pactworld (2018) Meet your miner: Prisca. Pact. https://www.pactworld.org/video/meet-your-miner-prisca. Accessed 06 Aug 2020
11. Bearor M (2018). Celebrating 20 years of WORTH. Pactworld. https://www.pactworld.org/blog/celebrating-20-years-worth. Accessed 06 Aug 2020
12. Pactworld (2018) Mines to market. Projects. Pact. https://www.pactworld.org/mines-markets/project. Accessed Aug 06 2020
13. IMPACT (2017) IMPACT in Tanzania. Transforming natural resource management Empowering communities. https://impacttransform.org/en/countries/tanzania/. Accessed 27 Jul 2020
14. Cartier LE, Ali SH, Krzemnicki MS (2018) Blockchain, Chain of custody and trace elements: an overview of tracking and traceability opportunities in the gem industry. J Gemmol 36(3):212–227. http://doi.org/10.15506/JoG.2018.36.3.212. 2018 The Gemmological Association of Great Britain. Available via Planet Gold. https://www.planetgold.org/sites/default/files/2019-12/Cartier%2C%20E.%20Laurent%3B%20Ali%2C%20H.%20Saleem%3B%20Krzemnicki%2C%20S.%20Michael.%202018.%20Blockchain%2C%20Chain%20of%20Custody%20and%20Trace%20Elements_An%20overview%20of%20tracking.pdf. Accessed 03 Jun 2020
15. Blore S, Ian S (2010) Taming the resource curse: implementing the ICGLR certification mechanism for conflict-prone minerals. In: International Conference on the Great Lakes Region(ICGLR). The Swiss Federal Department of Foreign Affairs and Partnership Africa Canada. ISBN: 1-897320-20-5. Impacttransform.org. https://impacttransform.org/wp-content/uploads/2017/09/2011-Mar-Taming-the-Resource-Curse-Implementing-the-ICGLR-Certification-Mechanism.pdf. Accessed 11 Aug 2020
16. Tanzanite Experience (2018) Your guide to buying Tanzanite jewelry online. The tanzanite experience. https://www.tanzaniteexperience.com/guide-buying-tanzanite-jewelry-online/ Accessed 12 Aug 2020
17. GemFair (2019) Debeers register. https://gemfair.com/. Accessed Jul 15 2020
18. Smith N.M (2012) Maasai and the Tanzanite trade: new facets of livelihood diversification in Northern Tanzania. Anthropology Graduate Theses & Dissertations. Paper 18. University of Colorado https://core.ac.uk/reader/54849188. Accessed Aug 05 2020
19. Poverty and Equality Databank (2017) Distribution of income or consumption by quintile. Available via The world bank. http://povertydata.worldbank.org/poverty/country/TZA. Accessed 25 Aug 2020

Jewelry Design in the Luxury Sector: Artistry, Craft, Technology and Sustainability

Veronica Manlow

Abstract The jewelry designer working within the luxury sector can be found in a variety of work environments. Within each of these contexts, ranging from the small private enterprise to the global luxury conglomerate, he or she will possess differing levels of control over and engagement in the creative process, design and making of jewelry as well as its promotion and sale. Jewelry designers who can be considered craftspersons/artisans and who may also be referred to as a designer maker may be found in all of these settings. Within the major luxury brands, only a few jewelry designers who work at the top level will have the autonomy to practice their craft in its entirety. The craftsperson represents the most deeply engaged category while at the other end of the spectrum, we find those who are involved in some but not all aspects of jewelry design. Many are subject to the logic of mass production which deskills jewelers and parcels out various functions to a variety of technicians in the interest of efficiency, scale and profit. A view of the work done by those engaged in jewelry design illustrates varying levels of skill, imagination, material consciousness and agency—or lack thereof when we consider contexts in which one's professional identity is compromised.

Keywords Jewelry design · Craftspersons · Luxury · Design · Jewelry · Materials · Innovation · Technology · Creative · Maker

1 Introduction

Jewelry designers within the luxury trade practice their craft in a variety of settings. Some may work for large global firms that are a part of a conglomerate, others may have their own businesses or work for mid-size and smaller firms that are publicly held or are private entities. The circumstances they work under constrain and enhance their creative process and shape their designs, the interactions they engage in and their

---

Veronica Manlow (✉)
Department of Business Management, Murray Koppelman School of Business, Brooklyn College, Brooklyn, USA
e-mail: veronica.manlow@gmail.com

I. Coste-Maniere and M. Á. Gardetti (eds.), *Sustainable Luxury and Jewelry*, Environmental Footprints and Eco-design of Products and Processes, https://doi.org/10.1007/978-981-16-2454-4_8

identities which inform the image of their brands. Learning from jewelry designers about how they practice their craft allows us an insight into the field and practice of jewelry design and making. It allows us to explore questions about the boundaries of craft and luxury. How are definitions of luxury being challenged by the emergence of new technologies and by business models and branding strategies which may obscure practices that are antithetical to craft and luxury?

One major distinction that shapes the jewelry designer's experience is whether they work within the realm of fine or costume jewelry, or in some combination of the two. Another is that designers may be in firms where the primary focus is fashion, or they may be in firms where jewelry is the sole focus. In a fashion firm such as Chanel, fine and costume jewelry plays a secondary role. In some fashion firms, jewelry is important but in terms of revenue it is not as important as leather goods which can bring in more than half of a brand's earnings. At Chanel, beauty products represent about one-third of the revenue [30]. For a brand like Chanel, its image is dependent on couture fashion and jewelry enhances that image considerably. Van Cleef & Arpels which sees itself as a "maison" specializes in jewelry as does Tiffany & Co. The latter has a much wider span of customers and had been in contract to be purchased by the LVMH conglomerate which at the time of writing attempted to renege on the deal. The creative process, work environment, conditions under which manufacturing takes place, the creations of jewelry designers and their own experiences within the luxury trade in large global firms differ from that of those who are independent entrepreneurs or who work for smaller brands.

We will explore four distinct experiences of jewelry designers which exist along a continuum in the luxury sector. The first two, embodied by Kari Woo and Mark Bloomfield, fall most clearly into the craftspersons/artisan model though one resists technology to a large degree and the other embraces it. I've had the opportunity to both interview and observe Bloomfield at work in his studio in Oxo Tower in London and to interview Woo remotely as well as listen to her speak about the work she does on a podcast. The third, Gina Ferranti whose brand is Gigi Ferranti, has many of these characteristics that align her with the craftsman artisan model but hers reflects the experience of an independent jewelry designer and entrepreneur who fully engages with the design process but whose jewelry is made by craftsmen with whom she has an ongoing relationship. This collaborative process where a jeweler works with others we will see presents to an even greater degree in the fourth type of experience represented by Queenie Cao. Cao's experience is reflective of a jewelry designer working for a fashion brand within the luxury industry who has much less connection to the craft aspect. Cao who worked for Marc Jacobs and who worked in the past for other luxury brands has recently launched her own costume jewelry brand. Both Ferranti and Cao are based in New York City. While Ferranti's jewelry is made in New York, Cao's designs for her own brand are fabricated in New York and in China. When she worked for Marc Jacobs, all jewelry was made in China with the exception of some runway pieces which were not for sale. As a means of comparison, I will also discuss jewelers and the technicians who work alongside them at the manufacturing plants of Tiffany & Co. Many jewelers work in hidden settings in jewelry stores, workshops, or factories. The number of those who work in factories are far greater when we look outside of the luxury industry but are substantial in luxury nonetheless as distribution increases and more product is sold to a wider audience. The experiences of those working at Tiffany & Co., in manufacturing plants across the USA, will be limited to an analysis of the posts they have placed on career websites as I have not had the opportunity to meet any of these people in person.

2 The Craftsperson

We will explore skill, imagination, material consciousness, desire to perform one's craft well and to be engaged in one's work in relation to the jewelry designers and jewelers we meet. Sennett [29] provides us with an important framework from which to begin considering the jewelry designers. It is one which can be extended to also include the social engagement, self and identity of the jewelry designer, something not considered by Sennett. In jewelry design, in particular not only the hand of the maker but also his or her persona which will encompass beliefs, philosophy and experience shape his or her company or brand. Adding these dimensions allows us to more fully consider the jewelry designer within a dynamic interactive social context characterized by ongoing interactions and dialogues rather than on the individual level alone.

Craftsmanship is a "way of life" that many say has "waned with the advent of industrial society." Richard Sennett [29: 19] sees it as an "enduring, basic human impulse, the desire to do a job well for its own sake." Sennett [29: 29] says that the craftsperson "occupies a special human condition of being *engaged*." In order to be worthy of the craftsperson designation, one must possess skill which once mastered becomes tacit knowledge. Sennett explains that this deeply engrained knowledge that one can take for granted is the foundation craftsmanship requires. One can excel in one's craft when tacit knowledge takes over and explicit awareness allows for judgments to be made. We might think of the example of the athlete, for example, an Olympic figure skater. He or she must have flexibility, strength training, aerobic training, etc. in addition to training in figure skating and dance. With intense practice, one begins to execute certain moves with ease. The skater cannot refine his or her moves and begin to develop artistry until a certain critical stage is reached. So too must the craftsperson repeat tasks many times to develop skill and to reach a state of mastery.

While skills "begin as bodily practices" for the craftsperson, "a technical understanding develops through the powers of imagination." The craftsperson will encounter challenges and problems that he or she will have to solve. Not having the right tool or not knowing how to achieve something has the advantage of calling forth creative thinking [29: 21].

Material consciousness is possessed by all craftspersons. He or she is "engaged in a continual dialogue with materials." The materials one uses are crucial to the creation of a final product. Knowledge of materials comes through experience. One can imagine new ways of using materials, for example, if one is motivated by concern for the environment or if one is inspired by advances in technology [29: 135].

3 The Jewelry Designer

Sennett [29] takes a broad view of the craftsperson. Musicians can be craftspersons as can potters or computer programmers, particularly those who use Linux. Jewelers certainly fall within these parameters whether or not they identify as craftspersons. A continuum exists and along this continuum we find many possibilities. While a craftsperson may also be a designer or a designer maker, a designer can be a distinct entity as can a craftsperson. A designer may conceive of a creation and plan its execution but not make his or her creation. A craftsperson may carry out the fabrication of an item based on a designer's creation. A craftsperson can make practical things with very little variation. Conversely, he or she can be seen as a master craftsperson or as an artisan: one who both designs and makes things that are unique and have artistic qualities. Some craftspersons may consider themselves artists, moving further into an aesthetic realm typically limited to painting, drawing and sculpture. When a jewelry designer is involved with both design and making, the full power of one's imagination is realized.

There are many roles that fall under the jeweler designation: master jewelers who have reached a pinnacle in their careers, bench jewelers who perform varied support tasks such as soldering and ring sizing, gemologists who grade diamonds and gemstones, and diamond and stone cutters and setters. And then there are jewelry designers. Jewelry designers "may or may not have the skills of a bench jeweler" but according to one jeweler, Jess Gebauer [8] "they should possess general knowledge of jewelry, fabrication methods, and stone setting styles." In addition to these roles, there are many support roles in the jewelry sector. This raises the question of how jewelry design is taught. According to one professor in a school of creative arts, a jewelry design curriculum required that students do the making and be proficient at bench work. He believes that jewelry design education and design education more generally have become far more superficial.

4 Luxury Jewelry Design

Kate Hubley [10: 48–49], a jewelry designer and gemologist, speaks of the ubiquity with which the word luxury is used in relation to jewelry. She believes that the "ownership of a particular artistic process" sets the "fine jeweler" apart from the "luxury jewelry designer." She asks jewelry designers, goldsmiths, jewelers and others what makes a piece of jewelry "true luxury." The definition she comes up with includes the use of noble materials, artistry and craftsmanship which include meticulous execution, intangible properties and a unique signature and longevity or timelessness. We will explore some of these features of luxury jewelry and will consider how luxury brands approach this in a way that differs from how they would be approached by a luxury jewelry designer who is not part of the world of the most recognized brands who simultaneously appeal the wealthiest of clientele and to a more general market.

Michelle Orman who works in public relations in the jewelry sector says: "indeed, using the noblest of materials is a top criterion, but luxury is much more than the monetary value of its components. Luxury is also artistry and craftsmanship." She then speaks of the craftsperson's work: "He or she will spend untold time—years in fact—honing and developing a proprietary technique that becomes their trademark and signature." Laurence Ravon who worked for Dior, Van Cleef & Arpels and Cartier (the latter two which are owned by Richemont) and now has her own atelier in Montreal believes that true luxury cannot be mass produced. She describes each stage as "meticulously executed." She worked on the Cartier Panthère among other pieces. Ravon has encountered many copies of the iconic Panthère and she says it is immediately recognizable to her as false by feel and by sight because it has not been made by a craftsperson trained by the brand. She can infer that quality and artistry are lacking.

Even luxury jewelers such as Cartier mass produce a great deal of their collections. For example, the LOVE collection which includes some more accessible pieces is a top seller. A yellow gold wedding band with the very recognizable love symbol surrounding the band can be purchased for $1,070. A pink gold, diamond and ceramic bracelet can be purchased for $1,210. While the upper end segment of fine jewelry is "volatile" says Jean-Christophe Babin, Chief Executive Officer at Bulgari and "One year it can grow by 50 percent and the following decline by 30 percent," Milena Lazazzera [21] who interviewed him says that more affordable jewelry "typically make up more than half of a house's sales and are a more consistent business." She contends that "semi-accessible precious jewelry is easy to craft" while high jewelry can take "thousands of hours to be completed and sometimes many years—especially when sourcing unique precious stones..." It also requires "a great deal of pomp and circumstance in order to close a sale." Bulgari (owned by LVMH) appealed to its higher end clientele by launching its Cinemagia collection to 200 clients on the Isle of Capri which had been closed off for the launch. It is not hard to see how luxury brands would capitalize on the publicity from their highest level creations (just as fashion houses do this with couture), to sell products on the lowest end. This is something in which LVMH has expertise and exploits.

Anne Sportun speaks not only of quality materials but also of something "intangible" which she describes as "immersive, experiential and sensorial." Hubley [10: 49] explains that Sportun's "signature" is unique and can be found not only in her jewelry but also in the "styling of her photos, her tone and manner, her product's packaging, and in her overall attention to detail." Branding allows for claims of unique design when it is not always merited. When we consider major jewelry brands such as Cartier, Bulgari or Van Cleef & Arpels, what we find is "perceived exclusivity" which Luca Solca [31] identifies as a "key to success." He states that "Leading luxury jewellery brands can get away with the highest possible price depth and can also serve a very broad constituency." The brand's "signature" that Hubley [10: 49] describes trickles down to its lower cost goods which cost much less to produce but which sell in great volume and still command high prices making it a "high gross margin business." Solca [31] says the business has a similar trajectory to handbags some years ago.

Ferranti, a jewelry designer we will meet shortly, is also interviewed and she adds another important dimension: "longevity." Considering what these practitioners have said, we find a definition that includes craftsmanship and artistry but that is further distinguished by a particular artistic process that distinguishes one jewelry designer's work from another's. This distinction includes intangible elements. A jeweler's signature is used in creating a brand identity. The materials one uses are exceptional. Luxury jewelry design has a timeless element which makes it worthy of being kept and passed down. These distinctions are interesting when we consider the connections jewelry has to fashion—a category which is very much of the moment and fleeting. How do the jewelry designers we will meet reconcile luxury and fashion and how do their tactics differ from those of the larger luxury brands?

5 Jewelry Design and Technology

Technology is an important feature in the creation and fabrication of jewelry. Jewelry designers use it at different stages of their work and to varying degrees in their creative process. Skill as said earlier is a trained practice. The jewelry industry has been transformed by the introduction of computer-aided design (CAD). In the past jewelry designers began with hand drawings which required advanced skills and experience in making jewelry and advanced knowledge about materials and techniques. David Weisberg [34: 21] explains how CAD revolutionized all professions in which design is central "In very simple terms, virtually no product, building, electronic component or system or factory is designed today… without the use of this." By the mid-1990s with the advancement of personal computers and a sharp decrease in cost, CAD use became widespread and no longer in the domain of highly specialized professionals within various fields.

The introduction of CAD allowed people who did not have an in-depth background in jewelry, or perhaps had no background at all, to design, 3D render and modify jewelry and to create wax models. Of course to do this well, a design background in most cases is a prerequisite. Joan Dalrymple [6] explains that this allowed for specialists in computer-aided manufacture (CAM) to emerge in the jewelry industry. She said that "in an ideal world the jeweler would be a CAM specialist with a formal jewellery education" and that this person would be "much in demand." The gates of course are also open for a system where people become highly specialized technicians and where fewer possess this "hybrid" training which gives them a full understanding of the capabilities of jewelry design.

This concerns Sennett [29: 62] greatly. He says that

> modern technology is abused when it deprives its users precisely of that repetitive, concrete, hands-on training. When the head and hand are separated, the result is mental impairment—an outcome particularly evident when a technology like CAD is used to efface the learning that occurs through drawing by hand.

At a time when CAD was just beginning to be more widely adopted by "craft/artist/studio jewelers," Sharon Elain Thompson [32: 30] raises pertinent questions about the role of the hand and the use of industrial equipment. She asks "If the hand no longer handles the object during its production, is the object handmade? And if industrial materials are used, requiring industrial equipment, is it a craft? The answers she receives from jewelers and educators are varied. For those who use CAD, these contradictions are not troubling, if they exist at all. Mary Lee Hu, Chair of the Metals Program at the School of Art of the University of Washington in Seattle and an artist/jeweler sees CAD and CAM as broadening the traditions of the University's metals program. She says that "far from removing the hand of the craftsman, I feel this is allowing our students to begin to do things not possible before" (2006: 32). Woo who taught extension classes at the Alberta College of Art and Design for a brief period said that CAD was not a part of the jewelry design curriculum. In that program "handwork is still the priority" she says though students can explore CAD on their own. Woo when presented with Thompson's [32: 33] questions sees a gray line and cannot say that the object is certainly handmade. I interviewed Woo in July 2020 and her views were similar. She could not have imagined how pervasive CAD would become. She has noticed a change compared to 10 or 12 years ago in the level of dexterity and agility of her students in the workshop classes she teaches. She believes that they have become less capable in this due to a reliance on computers. Woo points to an article in the Guardian where a similar problem was observed among young surgeons. Professor of surgical education at the Imperial College London, Roger Kneebone, points to a "decline on creative subjects at school and practical hobbies at home" and the emergence of smartphones as a reason for medical students and trainee surgeons not feeling comfortable doing things with their hands over the last 5 or 10 years.

> "We are talking about the ability to do things with your hands, with tools, cutting things out and putting things together ... which is really important in order to do the right thing either with operations, or with experiments. You need to understand how hard you can pull things before you do damage to them or how quickly you can do things with them before they change in some way."
>
> Kneebone said that by spending time online, children were missing out on practical skills acquired from hobbies such as cooking, woodwork, playing a musical instrument or model making. He endorsed making Halloween jack-o'-lanterns as a start for budding surgeons [33].

6 Sustainability and Ethics

Social and environmental responsibility in the design and manufacture of jewelry is often considered last, if at all, even if most brands proclaim that they are sustainable, ecological and socially responsible. We will consider the degree to which the jewelry designers, if concerned with sustainability, incorporate it into their work and business practices. Bloomfield and Woo highlight it on their websites and integrate it into the design and fabrication process. Certainly, when we consider the major luxury brands, we find that all have sustainability strategies but there are limitations to these strategies. The jewelry industry is notoriously lacking in transparency when it comes to its supply chain. This includes not only dealing with unscrupulous entities but also creating the conditions under which abuses were made possible. Human Rights Watch in a 2018 report states

> For millions of workers, gold and diamond mining is an important source of income. But the conditions under which gold and diamonds are mined can be brutal. Children have been injured and killed when working in small-scale gold or diamond mining pits. Indigenous peoples and other local residents near mines have been forcibly displaced. In war, civilians have suffered enormously as abusive armed groups have enriched themselves by exploiting gold and diamonds. Mines have polluted waterways and soil with toxic chemicals, harming the health and livelihoods of whole communities.

While no companies were excellent in sourcing policies and practices, Human Rights Watch considers Tiffany & Co. to be strong and Cartier and Bulgari to be moderate. Tiffany & Co. only buys from countries certified by the Kimberly Process Certification Scheme (KPCS). The KPCS which is focused on conflict is limited according to Human Rights Watch. It does not include any track and trace technology. The Maendeleo Diamond Standards go further in that they address: "labor conditions, child labor, health and safety, and environmental protection" and conduct random independent audits at mining sites.

7 Introduction to the Jewelry Designers

The jewelry designers that I consider can be categorized in terms of their level of engagement in all phases of jewelry design as having low to high levels of agency and command over their work.

Level of Agency/Command Over Work

Jewelry designers	Conception and design	Making	Engagement with customers	Selling and promotion	Agency/command over work	Professional satisfaction
Kari Woo	High	High	High	High	High	High
Mark Bloomfield	High	High	High	High	High	High
Gina Ferranti	High	Medium	High	High	Medium-high	High

(continued)

(continued)

Jewelry designers	Conception and design	Making	Engagement with customers	Selling and promotion	Agency/command over work	Professional satisfaction
Queenie Cao @ own brand	High	Low	Medium	High	Medium	High
Queenie Cao @ Marc Jacobs	Low	Low	None	None	Low	Medium-high
Tiffany & Co.	Low	Low	None	None	Low	Low

Both Woo and Bloomfield are personally involved in every aspect of the creative process including the promotion and selling of their designs. Neither one relies on others for any of the major decisions made in all processes in jewelry design. Gina Ferranti relies on other jewelers with whom she works closely in the technical aspects of design and the actual making of the jewelry designs. Her jewelry is mainly sold at jewelry stores so at a point she is no longer in complete control of the promotion and selling process. Cao's profile as the founder of her own brand is similar to Ferranti's except that she is not personally connected to those who realize her designs. When she worked at Marc Jacobs, her profile was similar to those working at Tiffany & Co. but in fact she enjoyed a higher status as a jewelry designer than those at Tiffany & Co. who worked under what could be seen as akin to factory conditions. In Cao's workplace, she enjoyed a high degree of satisfaction in her work even with low to medium levels of agency and command over her work.

8 Kari Woo

Woo is based in Alberta, Canada, and sees herself as an artist and designer maker. She is a strong advocate of the local and handmade economy. She is committed to working by hand in the design and making process. She has a craft degree in fine arts from the Alberta University of the Arts (formerly Alberta College of Art and Design) in jewelry and metals. She works with several assistants and if necessary will bring in another person if a project requires work in which she does not have expertise. Her creative process involves sketching by hand. She has looked at many items of jewelry designed on CAD and feels that they don't have the mark of the hand. The angles she believes are too perfect. She gives an example of the pyramids. With all the technology we have today, we cannot imagine how they were made. They were made because the time was taken to work out problems. Innovative solutions were found over many decades during which time people worked and constantly improved what they did gaining further expertise. They made mistakes and it may be through this process she believes that they found unexpected solutions to unsurmountable problems. Her most creative work she said has come from errors she made. These errors led to new ways of approaching her craft.

Woo has had many people suggest to her ways of expanding her business which often involved sending work overseas. She said this idea conflicts with her values on many fronts. She is committed to staying within a local small batch production model and working closely with clients. She believes that commercial luxury brands have coopted craft values and she cites one of her mentors, author and public speaker Dr. Sandra Alfoldy. Woo is a member of Citizens of Craft. The manifesto of the Canadian organization has ten points which reflect her philosophy: 1. You Are Not a Lemming; 2. We Value the Unique and Enduring; 3. Objects Should Inhabit, Not Intrude; 4. You are Not Automated, Manufactured or Cloned; 5. You Believe in 10-Digit Technology; 6. One Size Should Not Fit All; 7. Nothing is Newer Than Tradition; 8. Cookie Cutter Doesn't Cut It; 9. Vases Are People Too; 10. While We All March to Different Drums, We Move Together. Number 5 is explained in this way: "You connect more strongly with things made by a pair of hands—the original 10-digit machine." And 9: Craft objects evoke their makers, letting you surround yourself with not just things, but personalities [4].

Woo speaks of two sources of inspiration in her work which she describes as "versatile, lightweight and functional." Patterns in nature and minimalist sculpture and architecture from the mid-twentieth century are two of her most important influences (Images 1 and 2).

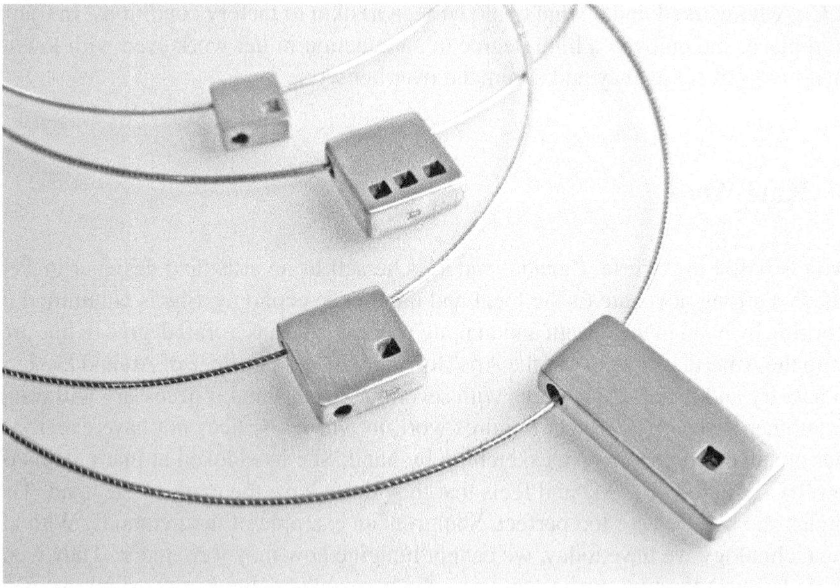

Image 1 Architecture series

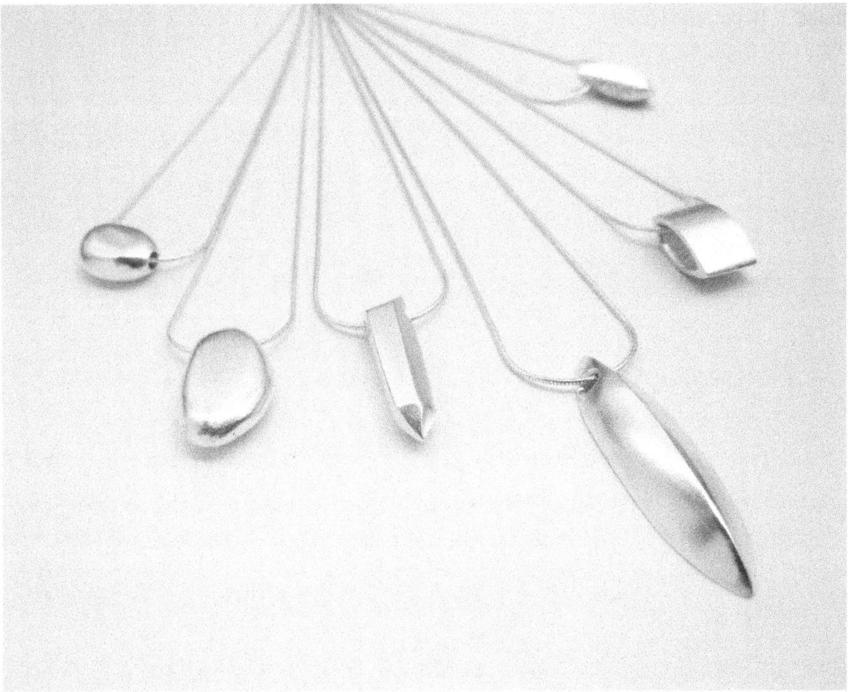

Image 2 Living matter series

Most of her pieces are limited edition, but she does not have any specific time frame and a collection can be available for a few years. About 65% of her work is commission based. She most values working with heirlooms as they are pieces that already have a story to them (Image 3).

On her website, we see many examples of engagement and wedding rings which were repurposed, and she says that many of her commissions are for wedding jewelry. Prices of jewelry available for purchase are quite reasonable in price. For example, the smaller ribbon hoops pictured below are CA $24. She said that this was a choice. Her jewelry is to be worn all day, every day and she wants it to be accessible. She states, however, that the prices are out of reach to many people. She never wanted to position herself as a high-end luxury brand with references to being "elite." (Image 4).

She describes her most loyal customer as someone of about 45 who is committed to buying local and to supporting artists and craftspersons. Whether a customer lives nearby or is an international tourist, she tends to share these values and enjoys knowing who made her jewelry and being able to buy directly from her. While she may be able to afford to do so, she would not buy from brand names such as Tiffany & Co. as a "matter of ethics, value statement or economics." Woo speaks of the appeal her aesthetic has to her customer

Image 3 Repurposed pearl ring

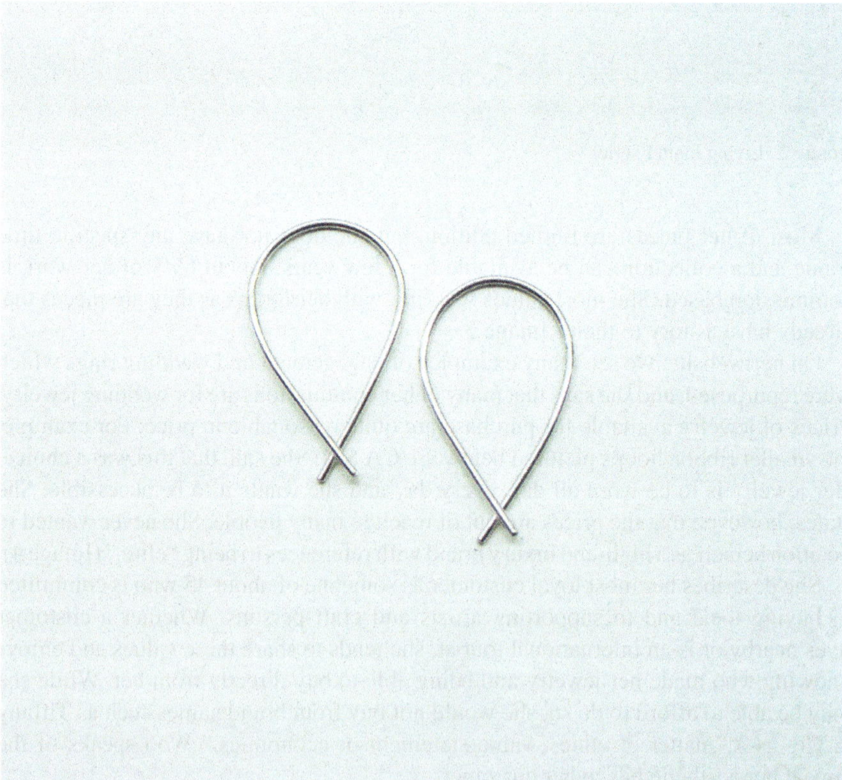

Image 4 Ribbon hoops, mini

> My aesthetic seems to speak to those who are not concerned with trends, but appreciate design that is timeless and distinct, minimalist and modern. Often the wearer of my jewelry designs has an understated sense of style; has an active lifestyle; travels often and appreciates versatile accessories that can transcend any outfit or ensemble. Young moms are often drawn to my work as it is low maintenance and easy to wear as well as athletes and career focused women who are always on the move.

9 Mark Bloomfield

Bloomfield refers to himself as an artist and also says that he is a designer maker. In addition to having a design and manufacturing studio from which he sells his jewelry, he is also the Visiting Professor of Design and Innovation in the School of Creative Arts at the University of Hertfordshire in the UK. Bloomfield considers his brand Electrobloom to be a sustainable luxury jewelry brand. He takes pride in combining traditional hand techniques with state-of-the-art 3D printing technology and is continually working toward further mastery in each area as it pertains to jewelry and design more generally. Having worked in large companies such as Asprey and Vivienne Westwood for 20 years, Bloomfield contrasts his approach with one which is focused on profit. He is not willing to sacrifice creativity and innovation in the interest of making more product, faster nor is he interested in this type of enterprise. As the founder of Electrobloom, he is able to determine a creative and business trajectory based on his values and commitments. This runs counter to the norms in large luxury brands where one is expected to present a self which is aligned with the brand.

His modular jewelry line can be reconfigured so that one item takes on various forms thereby encouraging less consumption. His use of recycled plastics promotes sustainability. Bloomfield says that the modular aspect came about as a means to create customized items of jewelry. Some customers seem apprehensive about it (Image 5).

He explains

> Although customers are made aware of this, few of them are willing to try it. They say "you are the designer, I want it as you intended it to be!" Some do have a go and together we will take items from the collection apart and then reassemble them into a new item(s) of jewelry that the customer has had a part in defining.

He goes on to say that while he originally came up with customizable jewelry as a means to involve the customer in the creative process, it had an unexpected benefit to his business.

> I no longer consider stock as fixed, it has become fluid. Through modular design I can take the jewelry apart and then reassemble so I can constantly keep the stock looking different. I don't have to clear unsold stock through a sale as I simply take it apart and reuse it. I can also introduce new component parts into the modular system and I have a whole new product line as existing components are recombined with new additions.

Bloomfield speaks of transforming single use plastics into luxury items, "turning waste into want." Milk, detergent and food containers are transformed into a material

Image 5 Secret garden long pendant necklace

that he says resembles semi-precious stone. Sustainability is built into his practice in terms of the materials he uses and his approach to design. Bloomfield also notes, "There's also a sustainable angle with being able to consider who the companies are that approach me with work, whether their business practice is one I agree with or not." (Image 6).

Image 6 Original flower group

Bloomfield spends his time “designing, thinking and making.” He draws inspiration as a jewelry designer from his work as an artist and vice versa. In collaboration with the University of Hertfordshire’s Digital Hack Lab where he co-invented the world’s first fully customizable 3D printed textile, he works on the design of 3D printed fashion garments which are customizable and utilize additive manufacturing techniques which make the supply chain transparent (Image 7).

His 3D sculpture Conform No 1–4, shortlisted for the prestigious Aesthetica Art Prize, has much in common with the flexible and interactive capacities which are a recurring feature in his jewelry designs. Drawing its inspiration from textiles, he describes this sculpture’s ability to enter into an “endless dialogue of expression and interpretation” implicating the user in the creative and design process rather than setting him or her apart from it as a separate and disembodied consumer.

I mention to Bloomfield that a person who worked for a large luxury firm in Paris says that artisans were seen as numbers, their production subject to various metrics,

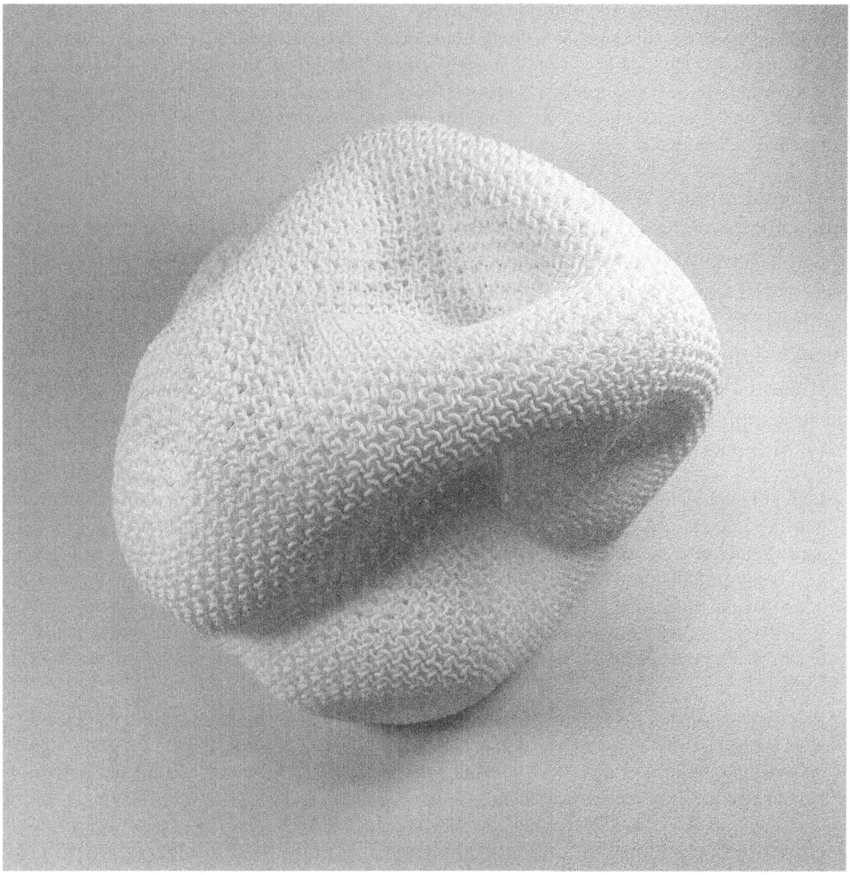

Image 7 Conform No 1–4

a process that she feels strips away their humanity and creativity. This undermines the notion of luxury which requires total dedication according to Bloomfield. He explains that he can take the time he needs to create one piece of jewelry. He sometimes works for weeks on custom pieces. He also devotes a considerable amount of time thinking about, researching and developing new materials and manufacturing processes. Many of his pieces are quite reasonably priced reflecting his commitment to a more democratic and accessible form of luxury.

Bloomfield speaks of his creative process. When he is designing for his own collection, he says

> I'll start with inspiration, a real flower for example or it could be a photo, a textile print or graphic, or something I've seen like the way architectural elements come together and the light catches it. Anything can spark an inspired moment, sometimes I'll just make a note of it on my phone. I'll also sketch, drawing is probably the best way of realizing an idea. I like to think of a blank sheet of paper as a realm of infinite possibilities. Once a mark has been made, I can then respond and lay other marks on the paper until it's doing what I want it to do although I may not of been aware of what I wanted until I started the sketch. Sketching is a process of introducing constraints into the creative process and allows me to quickly work things out or discover something unexpected. This initial idea then requires further refinement and I'll look to inform that development through questions like, how big should it be, does it fit the body, how heavy is it, can the customer use it easily, how much is it going to cost, are the sections of material strong enough, what color, surface finish, can it be easily finished/polished, etc.

When designing a bespoke piece for an individual client or when he is working with a company, he says that the process is largely the same. However, he says the questions have to be commercially oriented. It would be pointless to design something that cost 10 s of thousands of pounds if the client has a much lower budget in mind. Similarly, he says that one has to be aware of a design within market constraints when designing one's own collections.

Bloomfield states that "Figuring out the appropriate production process, materials to use, suppliers to involve, etc. is all part of the problem solving design process."

Of his general approach, he says

> I tend to approach each project differently as I like variety in the work I do. It helps maintain my interest and enthusiasm as I couldn't think of anything more demoralizing than approaching new projects in the same way over and over again. Although saying that I do also enjoy production work as I find repetition is the best way to hone skills and it gives head space to think. Also the opportunity to consider the way you're working becomes more evident as I'm constantly thinking about the way I'm sitting or standing, are things within reach, in the best position. Can I reorganise or reconfigure to achieve greater efficiency or comfort?
>
> Plus each project has a different set of requirements, different personalities in the customers I work with and different outcomes that may not be evident at the beginning, it's only through undertaking the project that these benefits come to light and become of value, not only to me but also the customers I work with.

Bloomfield explains that dialogue is a very important part of the creative process, particularly when he is working on one of a kind pieces with clients.

> The conversations are generally around the work to move it forward to a conclusion but also social chat and sitting down with a cup of tea are equally important. If I go to a meeting with a customer and I'm not offered a drink, it feels like something is missing.

When asked how his work now as an independent jewelry designer differs from work he did in fashion-oriented firms that make jewelry, he said

> There's a distinct difference between working for yourself and working as part of a team. everyone in the team needs to know what's going on and be able to run with the project to keep everything moving forward, even if they have to take a week off work, business continuity is a skill in its own right and is a challenge to implement and maintain. Working for myself I don't need to manage my work in that way, I have for example organized production in the UK when on holiday in Australia that would be ready on my return. Having this flexibility also adds an important sustainable quality to work, as through flexibility I'm able to sustain my interest and enthusiasm.

Bloomfield explains the technical side of the work that many designers do in larger luxury firms

> I have also done a lot of specification drawings for factories. It can be a challenge to approach a design, something you're familiar with, and lay out a blue print that describes that item in detail to someone on the other side of the world! It's an exercise in establishing a 2D visual language, have I captured everything needed to ensure I'm going to get what I want? I used to go through designs which my design team had produced with a fine comb, obsessing over detail to ensure they were communicating correctly. Waiting for a factory sample can take up to a month, you want to be sure that what you receive is what you asked for, otherwise you go through the process again, and again until you get it right.

He explains that in some fashion firms, designers may be closer to "stylists" than they are to jewelry designers.

> Stylists have the idea and then get others to make it happen for them. It is important to at least have some understanding of the materials, production processes and techniques as this enriches the designer's pool of expertise and allows them to intelligently bend the rules and shift expectations from an informed position. New technologies are still being investigated by many different companies and individuals with different agendas, all resulting in different outcomes. Even the reinterpretation of traditional techniques is still being undertaken in order for people to expand their understanding and show new ways of achieving things. I think that's the point.

Bloomfield uses a variety of traditional hand techniques in combination with advanced digital processes such as 3D printing technology. He sees technology as a tool, and he does not see a contradiction between craft and technology so long as the person using the "machine" has an understanding of what it does.

> Craft generally requires the use of tools. Even if you're braiding together reeds you'll use your fingers which are tools! Industrial equipment tends to be a refinement of those hand processes. You can for example build a forge to heat metal to cast that molten metal into cavities made from sand or cuttlefish bone, which are techniques that are still in use, or you can use the lost wax process, where the wax is 3D printed, and the casting machine is state of the art. You need to have the skills, experience and understanding of the casting process in order to get the best out of it regardless of the equipment used. Designing something, a ring for example without understanding the method used for manufacturing can be a bit hit

> and miss. The designer wants the ring shanks to be that thin, the casters says it won't cast, the designer says do it anyway, the casting fails.
>
> Often the background to all the industrial processes are hidden, people just see the machine and the button, press the button and out pops the thing. In order to make that happen an extraordinary amount of know-how, skill and understanding has been invested to make that happen. When someone new has to operate the machine, they also need to understand what is happening, how the material is being manipulated, what the machine actually does etc.

Bloomfield explains that he has been spending lockdown during COVID-19 "getting to grips with my software as it's undergone many updates or there's functionally that I don't really need to use in my day to day work." This has taken a lot of his time and attention. He speaks of procedurally-driven materials and says

> I understand what it means but the number of math driven options and the various ways the chain of instructions can be linked together means there's thousands of potential outcomes. Some of the sliders and number boxes can take 0.001 as an input, I'm enjoying the detail and starting to recognize the difference between these tiny adjustments. How do you animate procedurally so that the output is slightly different each time?

He says that in part, this is "playful exercise" and it is fun but he asks himself "how do I sustain it?" He is constantly thinking about what he can do with this new knowledge and with the skills he is building. Bloomfield says that he is starting to think of an entirely new project which involves jewelry as wearable screens.

10 Gina Ferranti

Gina Ferranti's brand is Gigi Ferranti. She chose her nickname for the brand because she felt it had more ambiance. She started out in the fashion industry, an experience she feels informs both her creative and business approaches. Her long-standing fascination with jewelry led her to enroll in the Gemological Institute of America (GIA) where she earned a degree in gemology in 2011. After working for David Yurman, she decided that she wanted to design herself. She went back to GIA and started their jewelry design program. Beth Bernstein who writes for Forbes named her as one of the "six jewelry designers that should be on your radar for 2019" and said that she belongs to a "new wave" of jewelry designers that "weave together old world techniques with modern technology."

Ferranti when speaking of jewelry speaks at length about stones, their properties and based on factors such as hardness and grade, how they can or can't be used and should be set.

> If you look at my work I tend to use a lot of diamonds and a lot of sapphires because they are the top two hardest minerals that are made on earth. I do use some garnets too. Diamonds being the hardest which is a 10. Sapphires being a 9, 9 and a half. Garnets are around 7 and a half. But if I am using something softer than that usually is bezel set which protects the stones. You have to know how to set the stones. You can have a beautiful chalcedony or cabochon but know how to protect it or use thicker prongs that come up on the sides.

She believes being a gemologist gives her an "edge" over jewelry designers without that education. We see how her knowledge allows her to create jewelry which can be seen as luxury.

> Being a gemologist gives me an edge because I know what to look for when purchasing a stone for a particular design. First is the hardness and durability of the gemstone- is it safe in the setting for wearability. For example emeralds have a lot of fractures in the stone and you want to make sure they are not reaching the surface because when the setter sets the rings he could crack it if he pushes too hard. I also know what is a precious stone compared to a semi-precious which has a lot to do with the value of a piece of jewelry, for example sapphire, ruby and emeralds are precious. Most other colored stones are semi-precious. Let's talk about quality– within each gem species there are grades of quality which has to do with the rareness of the stone. What makes it rare is how clean of inclusions it is, color saturation, cut of the stone. All these factors add to the quality of the stone. Macy's sells sapphires for very little, they are considered commercial goods. A quality I would never use, hence the price difference. The rareness of the stone produced in nature also makes it more expensive and desirable. Diamonds for instance come in many different qualities. The whiter the diamond with less inclusions is much more expensive because they are more rare to find in nature. My education and learning never ends. You also have to stay on top of current trends with certain gemstones and colors just like fashion.

When asked about sourcing of stones, she says

> I don't go to the mine and buy rough stones but there is the Kimberly process which keeps the supply chain honest. All the people I use are a part of the American Gemological Trade Association. There are check and balances. But it is impossible to always 100% know. The industry is focusing and we are getting better at it and I always am willing to see something. If I can know what mine it came from I always put it down. And if I knew it came from a disreputable source I wouldn't buy it.

Ferranti believes that the "hands on" experience of actually making the piece and making lots of mistakes and errors is where the real expertise comes in. In doing this, she says you also discover something new. She spoke of her grandfather who built roads and sidewalks, and her father who built buildings. She sees a continuity in the work that they do and what she is doing.

> So my dad would start with a sidewalk and see miles and miles of the work where he started from. It is very rewarding. It is kind of the same parallel for me when I started design. And I see from start to finish how it ended up and how it finished. It is the whole process that is very rewarding.

She explains how her creative process begins, saying that she thinks very architecturally.

> I can be out driving and I see a motif, a window or even a door. A portion of an architectural element. It doesn't have to be a whole style or something. My brain doesn't stop. I keep a pad by my bed. And you can't make everything all the time so you hold all those little sketches and eventually it comes into play.

She explains that her Lucia collection which is her largest one reflects this thinking (Images 8 and 9).

Image 8 Lucia stackable rings

Image 9 Lucia 5 station necklace with carnelian inlay

> Lucia means light in Italian. It kind of mimics stain glass windows. And that kind of follows through with other pieces. That is the concept of Lucia. You can stack them. You can buy one and you can keep collecting them. It changes the whole look.

Ferranti explains the phases in her creative process. While conception and execution of the design are carried out by her exclusively, she turns to another designer to translate her drawing into a 3D simulated design.

> First comes the inspirations, and if you remember my former career was in fashion, so I do picture the women dressed and wearing the pieces. Then sketch down on paper several drafts and then finalize. From there I bring it to my CAD designer who puts it into a computer-automated program which is in 3d. We then discuss how it will function if there is a moving part etc. From there we print a 3d wax and try it on to make sure its comfortable and wearable as well as esthetically beautiful. The wax is then cast into a sterling silver model of the item and then a rubber mold is made. Each piece is then cast from the rubber mold.

It is at this point that Ferranti will have the jewelers she works with fabricate the jewelry using the rubber mold. She keeps the molds and can always decide to make more pieces.

Image 10 Cleopatra ring

Ferranti has won two awards for her designs. One was the Gem DIVA award from the American Gem Trade Association (AGTA) for her Cleopatra ring (Image 10).

11 Queenie Cao

Queenie Cao studied fashion at the China Central Academy of Fine Arts and after that at The Glasgow School of Art and at Parsons in New York. She began a career in fashion design before transitioning into jewelry design. Cao worked as an assistant jewelry designer at Marc Jacobs. She is currently working as a freelance jewelry designer and has launched her own company Queenie Cao Jewelry where she sells her designs online and through a variety of retailers in New York and Tokyo.

When at Marc Jacobs, she worked in a team consisting of her and a senior jewelry designer. They both worked with a creative director. Ideas tended to come from the creative director or from Marc Jacobs directly, but there was a collaborative process. Cao's experience at Marc Jacobs was an experience entirely different from the work experience of Bloomfield or Ferranti. When asked about how the creative process worked at Marc Jacobs and what her role was as a jewelry designer, she explained it in this way:

> I think from my experience working in the firm, it's like a team work. It's not your personal creative, it's a team creative. We come up with ideas together and then research and develop. It will represent the work of the company not yourself.

I asked her about how she felt about the work that she did and if she felt she could be very imaginative within the context of the team.

> Yes, of course, if you do the job you like. We come up with ideas together and research. It's very imaginative, but like I said, it's not your own work. Sometimes the creative director will have the idea and you will design and create based on that.

When asked about her ideas and if she felt free to communicate and express them, she explained that she had less ability to do so than others. I mentioned to her that other jewelry designers had said that at larger luxury firms the design process could be quite regimented. To this, she responded: "Yeah I think it depends on the company size and team size. If it's a big team it will hard to be creative if you are not in a high position." Marc Jacobs is quite small and so there is only one team of two. In larger companies, one may have several teams.

She expressed liking teamwork better than working alone but said that she liked both types of work. A lot depends on the company and its culture. She enjoyed working with the team and liked that the pace was fast. "We had to work fast and so it will be a lot in a day that we did." Sometimes, they worked on different seasons at the same time and runway season coincided with a commercial season. She explained that had this been a larger fashion firm, the model would have been more like that of fashion design: "it will be more people and not everyone can get creative unless you are in a high position."

Cao explains how the creative process works

> We search for and prepare the mood board and develop. They will choose what they like. The senior designer has a more creative part in this. I do more of the technique part.

CAD is a major focus for Cao. A typical day for Cao consisted of doing tech packs and performing the necessary research to develop them. She said that in this way her work is very much like fashion design with her working most of the day on a computer program. "There is no actual piece making in the company" says Cao.

Would Cao like to make jewelry on her own? She says that she would love to, but she does not have any tools or equipment.

She has designed several collections in her own firm. The Amor collection is her favorite, to which she would like to add. Of this collection, when asked about how she came up with an idea for it, she said that she wants to view the world with love.

We see in Cao a desire to engage more deeply in her craft than she did perhaps in the past at Marc Jacobs (Images 11 and 12).

> For the jewelry collection of Amor, I would like to develop new possibilities of wearing jewelry and see modern jewelry in different ways. Using different types of materials and

Image 11 Amor double heart drop earrings, large

types of jewelry. What jewelry could offer us is not only its beauty but also the excitement and joyful offer of wearing it.

Indeed, we can see in this particular design a lot of joyful optimism.

On her website in the opening page, one sees the words LOVE is LOVE and underneath that #PRIDE. I ask her if she has a particular connection to the LGBTQ community or if this was somehow an influence in her design of the Amor or Love+ collection.

LGBT for me is important. It's like my friend circle. Not related to design specifically but it's an important community in fashion.

We see that she approaches her design process from a fashion sensibility (Image 13).

Amor Eyewear is Cao's favorite design. Brazilian drag queen and pop star Pabllo Vittar and Latin Grammy winner Bad Bunny have been photographed wearing this

Image 12 Amor earcuff

unique piece featuring hearts and diamonds priced at $875. It is pictured below on a model (Image 14).

Cao gives a very straightforward and honest answer when asked about sustainability concerns: "I don't know it's very hard to be sustainable in fashion to be honest." She goes on to say that "Companies all have policies but in truth they are not really sustainable."

12 Tiffany & Co.

Tiffany & Co. is headquartered in New York. Founded in 1837, it grew into a world renowned jeweler. Reflecting on the work of his mentor Hannah Arendt, Sennett refers to "homo laborans" in her work "a drudge condemned to routine" while "homo faber" is "the man as maker" ([29]: 17). For Arendt [1: 83], labor work and action are the three fundamental human activities. She investigates the ways in which labor has been viewed from antiquity to the present. For much of history "to labor" meant to be "enslaved by necessity." In modern theory, we find that a contempt for labor gives way to "a glorification of labor as the source of all values." [1: 85]. For Arendt unlike

Image 13 Queenie Cao assorted designs

Karl Marx, there is a hierarchy within labor and animal laborans remains in the animal realm while for Marx, the laborer is elevated to the position of animal rationale. The maker unlike the laborer for Arendt, through his or her work, participates in the collective and even political life or his or her community. Action and contemplation for Arendt are higher forms of being. Whether we think of the jewelers working for Tiffany & Co. as homo laborans or as Marx might see them as a homo faber and animal rationale who has a fundamental dignity even if he or she may be enslaved in the current economic circumstance, what we find at Tiffany & Co. is a jeweler stripped of his or her agency and creativity. In this context we find a jeweler alienated from his or her products and from his or her handwork.

Tiffany & Co. has 7 manufacturing sites in the USA where jewelry is made: 3 in New York (Pelham, Mt. Vernon and East Elmhurst in Queens), 2 in New Jersey (Parsippany and Hanover) and 1 each in Cumberland, Rhode Island and Lexington, Kentucky. Finished stones are sent from the company's factories in Africa, Europe and Asia. The two facilities in New York in Pelham and in Kentucky produce 70% of the jewelry sold globally [27]. There were 125 employees at the time the Kentucky plant was founded in 2011, and there were almost 800 in Rhode Island, 500 in Parsippany, 200 in Hanover and 100 in East Elmhurst in 2015 [7, 27].

Image 14 Amor eyewear

> The largest factory [in New York] is located in Pelham, New York, and produces everything from solitaire diamond rings to diamond earrings, necklaces, and bracelets. Its satellite factory in Mt. Vernon, New York, handles overflow from Pelham.

In a job posting for a jeweler in Pelham, we see how Tiffany & Co. describes this location and the work that is to be performed. "The Tiffany Workshop is where our artisan/master jewelers bring new designs and Tiffany archival sketches to life many for the first time." Ten years or more of experience is required for this job where one is expected to possess "Ability to fabricate jewelry and/or jewelry components from concept to finished product using sketches, renderings and CAD designs." [23].

There are 7 employee reviews on the website Glassdoor for the Pelham location and there are 3 on Indeed. Eight out of 10 are bad and there is mention of racism, bodily injury and intimidation.

A gemologist and a manufacturing employee mentions racism. The gemologist [9] says that a "robust EOE program" is touted but in fact he or she does not see minority employees promoted beyond the "junior management bracket." The manufacturing employee [24] says that "100% discrimination and racism is practiced openly."

A model maker [26] states that there "Seems to have been a consorted [*sic*] effort to discriminate legally against injured workers, some are fearful of reporting injuries."

Apart from these more serious charges, there are complaints of low pay, poor communication, lack of advancement and poor treatment.

The manufacturing employee [24] says that there is "absolutely no recognition or appreciation for excellent work." He or she states that

> Employees are considered a NUMBER and not a HUMAN BEING. A revolving door and an absolute disgrace to any work-force. The few that stood up to make changes were forced out and it continues.

This person praises his or her coworkers who help each other "get through the day without losing their minds."

A coordinator [5] describes the plant as a "slave driving hell" where there is no concern for health and working days exceed 12 hours.

In East Elmhurst, Queens, we have 2 bad reviews. A jeweler [18] who was there more than 10 years speaks of "no challenge" and management not caring "anymore." A jeweler [14, 15] who is a woman who had been there for more than 8 years points to favoritism of male employees and says: "Managers make rude and sexist remarks in the workshops. Was told it is just 'locker room talk' and 'no big deal.'"

On a "super-duper day" in the Blue Grass Business Park, the Mayor of Lexington, Kentucky, welcomed Tiffany & Co. officials on the inauguration of its "25,000 square foot manufacturing plant" [7].

> The facility in Lexington, Kentucky, is a relatively new venture and primarily produces solitaire diamond rings with a small assortment of other jewelry. … All merchandise, once completed, is shipped via armored carrier to the company distribution center in Parsippany, New Jersey. [27].

All 8 reviews at this plant are bad. Problems include low pay, long hours, high pressure, no room for advancement, little recognition, poor management, injury and abusive conditions. A bench jeweler [2] who worked for the company for more than 3 years gives some advice to management "Appreciate your manufacturing staff, if not for us, you'd be nothing." He or she speaks of "mandatory overtime every weekend, repetitive products, no bonuses" and yearly raises being "dismal."

A jeweler/engraver [20] who was at the company for more than 8 years describes pay and benefits as "cataclysmically poor" and explains that this is coupled with "high expectations" yet jewelers have "low quality materials and designs" to work with.

A jeweler assembly/polisher [19] says:

> I enjoy manufacturing jewelry but there is a big lack in communication within the facility. The pressure to keep efficiency up is sometimes unrealistic. They want to be efficient but when we are faced with a new problem to solve or are attempting to be efficient at other aspects of our jobs, we get push back or we are lacking tools.

A recurring theme, a jeweler [13] who has been at the plant for almost 6 years feels disregarded and abused and he criticizes the quality of the jewelry they are manufacturing:

> Management is terrible! No one cares about you, you're just a number, they're abusive and will terminate you with no warning if they need to. Everyone is injured and they send you to a subpar PT and expect you to move on. They force you to work OT, lie to your face, and use you for anything they can. DO NOT work here. This place is a joke and they're producing terrible jewelry that has holes, is hollow, and doesn't meet standards.

Another jeweler [16] also speaks of increasing pressure and decreasing quality in the face of high rates of injury.

> They churn out an ever-increasing number of pieces with ever-diminishing quality of stones and castings. Well over 50% of the workers in the facility have a work-related injury with no changes to remedy the conditions. Workers comp is handled by an outright criminally negligent, or just plain fraudulent vendor.

He or she says "The only people Tiffany & Co care less about than their employees are their customers."

A jeweler [18] also speaks of a workload that is too demanding:

> Unreasonable demands for workload, with the threat of write-ups and termination if requirements are not met. While the coworkers are good people the anger and stress levels are so high that it is a miserable place to work. Management doesn't know what the other hand is doing. Every day they are trying some new idea on us to see if it works.

The largest factory in Cumberland, Rhode Island, produces all the metal jewelry "from silver bracelets and earrings to platinum and gold settings for diamond engagement rings."

> The Cumberland facility is also home to the hollowware shop, which produces silver platters and dinnerware and crafts most of the professional sports trophies awarded across the United States.
>
> Silver merchandise, a small but important part of the supply chain, is produced in the Rhode Island factory and then shipped to a Dominican Republic factory, which provides the final polish for all silver jewelry sold at Tiffany & Co. Once the product has reached a level of polish that meets Tiffany standards, it is sent back to Rhode Island for final manufacturing [27].

Three jewelers at this location have positive reviews and are satisfied with their pay and the benefits they receive. One jeweler [17] who worked at the location for more than 10 years says "good pay but could be better" but feels management could learn more about the jobs the jewelers do. The second jeweler [13] has no complaints and cites "health insurance, paid holidays, dental, vision and discounts" as positive points. The third jeweler [14, 15] described loving his or her job. He or she describes Tiffany & Co. as a "very strict ruled company."

> Tiffany is a good place to work. My responsibilities were to clean casting of platinum jewelry, gold and silver. Each piece is precisely cleaned to perfection. To fish remove tabs from wax moldings and castings. And then to removing all flaws from surface. Some pieces were a little complicated they were assembled and soldered then removed excess metal from rings. All processing is done with electric fortem with different pieces to be used accordingly. When cleaned then sent to wash and then polish for the beautiful flawless finish.

Another very positive review comes from a Lean Six Sigma Black Belt [22] who worked in Culture Development and Training.

A jeweler [12] expresses his discontent with the structure of the work environment and gives us some insight into this structure to which he objects:

> This company lures you in with a higher than usual hourly rate, then suck[s] all the life out of you trying to meet unreasonable production standards. I'm not a robot, I've been a jeweler my adult life and there may be maybe 2 dozen actual Jewelers in that building. Everyone else is a glorified one task robot. NO room to advance to where my knowledge would be of use. Not to mention the HR department is a joke. Unless you are a brown-noser you will go nowhere here. The atmosphere is not a safe one and injuries are abundant. Stay away.

In New Jersey, jewelry is hand-finished in Parsippany. In Hanover, "engraving, hand engraving, and etching" are performed.

We hear from a polishing/machine engraving specialist [28] who was employed at this plant for 18 years:

> When I got my job there in 1999, it was a[n] excellent place to work. I use[d] to just love going to work not because I am being paid, but enjoy the work environment. New management took over and the place just went to the dogs. They got rid of all good employees that were dedicated to their jobs and go above and beyond to get the job done especially people that have been there for many many years. They wanted to save money so they got rid of the people that invested all of their lives to that company, because they were over 62 and making a lot of money for their years of service, and they got new employees that don't care about the job as people [who had] been there for years.

13 Conclusion

While we have only had the narrowest of views into the professional lives of the jewelers and technicians at Tiffany & Co. and no access to the few high-status jewelry designers who work in the company, we see that those who have spoken out are not able to use their imaginations or to achieve the level of agency we see in the 4 jewelry designers that were interviewed. All jewelry designers must work under conditions where constraints are ever-present but for the jewelers at Tiffany & Co., both their individual and social capacities have been severely curtailed. For Cao, when she was at Marc Jacobs, we see certain parallels in that the creative potential of her work was limited under hierarchical conditions. However, she was an active member of a team and felt that her contributions were valuable. Furthermore, she enjoyed being in a fashion environment where there was variation and change and where her skills were appreciated.

Bloomfield, Woo and Ferranti have great deal of freedom in that they can each determine the types of collections they make. While Bloomfield and Woo are connected to each part of the design and making process, Ferranti passes some aspects of design and making to others: her CAD designer and the jewelers who complete the fabrication of her designs, yet she remains connected to the craft process through the relationships she has with the jewelers who fabricate her designs. Cao has now joined the ranks of these 3 jewelry designers in starting her own jewelry brand of which she has full creative control. We see that her relationship with the creative process ends at the point of fabrication when work is sent abroad or in some cases is made in New York. In this way, she is following the pattern familiar to her within the fashion sector. Her training is in fashion and not jewelry design so unlike Woo, she does not have the training in metals nor the expertise that Ferranti has in gemology. We do not see the same level of engagement that we see with Bloomfield in mastering new technologies. For Cao, her satisfaction comes through design and styling and this informs her approach to her business which is based on a fashion industry model. Woo in her commitment to craft is on the other end of the spectrum as an artist and maker.

One's thoughts, attitudes and ideas, indeed one's imagination for George Herbert Mead can be understood by the actions one takes. Says Mead [25: 10] of consciousness and mental phenomena: "If we then conceive of it functionally and as a natural rather than a transcendental phenomenon, it becomes possible to deal with it in behavioristic terms." In giving a central position to acts in human conduct, Mead adds an important dimension of sociality. An individual possesses an "I" which initiates novel responses and a "me" which represents a generalized other. In interactions, the attitudes of others change and so too does the me. This is what makes the self a dynamic entity.

At Tiffany & Co., we see a situation where the I is extinguished and the me is shaped by a generalized other who is seen as an adversary and indeed as one who wishes to extract as much labor as possible with no regard for enhancing the professional identity of the jewelers. This is no doubt the case for many involved in jewelry design within the luxury fashion context or for luxury jewelry brands who hold positions far removed from the brand's corporate offices.

In the case of the independent jewelry designers, his or her I may be more prominent at times when he or she decides to take decisions that are not in direct consultation with others. There is a capacity in each of these individuals for an independent direction which is motivated by his or her values and commitments whether they may be centered around local production and a deep engagement with the materials with which they work as we have seen with Woo or to expanding ones horizons through the mastery of technology as we have seen with Bloomfield. Ferranti and Cao are more directly connected to the world of fashion. Fashion and a concern with craft are not mutually exclusive. We see in Ferranti's case that fashion informs her designs and choice of gems used. Ferranti is able to connect her knowledge of gems to the colors that are on trend during a particular season bringing this sensibility into her collections. Cao engages more directly with popular culture. She works with broad concepts such as love which she connects to communities and social movements in a way that resonates with a younger demographic.

For jewelry designers, reflection is linked to an internal dialogue and to an engagement with others. The designs they put forth for their clientele and for the larger public whom they hope will accept their creations as meaningful objects require an ability to take the role of the other. In Cao's case, that other might have been Marc Jacobs when she worked at this firm—even if she had little or no contact with him. She was required to embody the aesthetic of his brand as conveyed to her by his intermediaries. It is critical for a jewelry designer who has full control of his or her creative abilities to be able to assume the attitude of his or her customer. The capacity to take the role of multiple others such as one's private clients (who each have their own profile) or a new demographic one wants to reach out to—while at the same time cultivating a unique identity—allows the jewelry designer to develop a persona that animates his or her brand and distinguishes it from those of competitors.

References

1. Arendt H (1958) The human condition. The University of Chicago Press, New York and Chicago
2. Bench Jeweler (2018) Bench jeweler. Lexington, KY, 8 Mar 2018. Available via Glassdoor https://www.glassdoor.com/Reviews/Employee-Review-Tiffany-and-Co-RVW19598800.htm. Accessed 15 June 2020
3. Bernstein B (2019) Six jewelry designers that should be on your radar for 2019, 17 Dec 2019. Available via Forbes https://www.forbes.com/sites/bethbernstein/2018/12/17/six-designers-that-should-be-on-your-radar-for-2019/#7a6258da1470. Accessed 1 May 2020
4. Citizens of Craft (2012) Manifesto. Available via https://citizensofcraft.ca/manifesto. Accessed 11 July 2020
5. Coordinator (2012) Slave driving hell. Pelham, NY, 5 Aug 2012. Available via Glassdoor https://www.glassdoor.com/Reviews/Employee-Review-Tiffany-and-Co-RVW1810892.htm. Accessed 17 June 2020
6. Dalrymple J (2010) CAD/CAM impact on jewellery fabrication. Available via Ganoksin https://www.ganoksin.com/article/cadcam-impact-jewellery-fabrication/. Accessed 20 Aug 2020
7. Fortune B (2011) Tiffany opens New Lexington plant that will employ 125, 28 Sept 2011. Available via Lexington Herald Leader https://www.kentucky.com/news/business/article44128140.html. Accessed 11 July 2020
8. Gebauer J (2020) A jeweler is a jeweler is a jeweler or so you thought. Available via Hammer and Gem http://hammerandgem.com/jeweler-jeweler-jeweleror-thought/. Accessed 17 July 2020
9. Gemologist (2016) Productive yet fear driven workplace. Pelham, NY, 29 Jan 2016. Available via Indeed.com https://www.indeed.com/cmp/Tiffany-&-Co./reviews/productive-yet-fear-driven-workplace?id=0a644ba71a139680. Accessed 19 May 2020
10. Hubley K (2019) The luxury experience: building a high-end brand. Jewellery Business, July 2019, pp 46–50
11. Human Rights Watch (2018) The hidden cost of jewelry. Human rights in supply chains and the responsibility of jewelry companies, 8 Feb 2018. Available via https://www.hrw.org/report/2018/02/08/hidden-cost-jewelry/human-rights-supply-chains-and-responsibility-jewelry. Accessed 21 July 2020
12. Jeweler (2020) Do not work here!!, Cumberland, RI, 5 Mar 2020. Available via Indeed https://www.indeed.com/cmp/Tiffany-&-Co./reviews/do-not-work-here?id=13f367753921a472. Accessed 11 Aug 2020
13. Jeweler (2019) Horrible. Lexington, KY, 23 Jan 2019. Available via Indeed https://www.indeed.com/cmp/Tiffany-&-Co./reviews/horrible?id=45544b93743da757. Accessed 29 July 2020
14. Jeweler (2018) I loved my job… I had relocated to Texas. Cumberland, RI, 10 Sept 2018. Available via https://www.indeed.com/cmp/Tiffany-&-Co./reviews/i-loved-my-job-i-had-relocated-to-texas?id=d9b883cab95022b7. Accessed 19 July 2020
15. Jeweler (2018) Do you love mansplaining and sexism? Then this is the place for you. East Elmhurst, NY, 11 Nov 2018. Available via Glassdoor https://www.glassdoor.com/Reviews/Employee-Review-Tiffany-and-Co-RVW23366238.htm. Accessed 14 May 2020

16. Jeweler (2017) Hands down the worst company I have ever worked for. Lexington, KY, 1 June 2017. Available via Indeed https://www.indeed.com/cmp/Tiffany-&-Co./reviews/hands-down-the-worst-company-i-have-ever-worked-for?id=0262d3d4e851c62f. Accessed 11 Apr 2020
17. Jeweler (2016) Jeweler. Cumberland, RI, 16 Jan 2016. Available via Glassdoor https://www.glassdoor.com/Reviews/Employee-Review-Tiffany-and-Co-RVW9219056.htm. Accessed 28 July 2020
18. Jeweler (2014) Very poorly organized. Lots of presser [*sic*]. Lexington, KY, 23 July 2014. Available via Indeed https://www.indeed.com/cmp/Tiffany-&-Co./reviews/very-poorly-organized-lots-of-presser?id=f80af6ad651c1943. Accessed 1 Aug 2020
19. Jeweler Assembly/Polisher (2020) Unrealistic expectations. Lexington, KY, 6 July 2020. Available via Indeed https://www.indeed.com/cmp/Tiffany-&-Co./reviews/unrealistic-expectations?id=472fbfbf0b1b6016. Accessed 6 July 2020
20. Jeweler/Engraver (2019) Worst company I ever worked for. Lexington, KY, 6 July 2019. Available via Glassdoor.com https://www.glassdoor.com/Reviews/Employee-Review-Tiffany-and-Co-RVW27828259.htm. Accessed 1 June 2020
21. Lazazzera M (2020) What's really driving success in the fine jewelry market? The business of fashion, 17 Jan 2020. Accessed via Business of Fashion https://www.businessoffashion.com/articles/professional/how-affordable-diamonds-became-luxury-jewellerys-cash-cow. Accessed 11 Sept 2020
22. Lean Six Sigma Master Black Belt (2020) Supportive work environment, 8 June 2020. Available via Indeed https://www.indeed.com/cmp/Tiffany-&-Co./reviews/supportive-work-environment?id=6eefd1658a7156d9. Accessed 11 July 2020
23. Lensa (2020) Jeweler TWS, 19 Mar 2020. Available via Lensa https://lensa.com/jeweler-tws-jobs/pelham/jd/91108e0bd0e39e96f32133ca811e59c2. Accessed 22 July 2020
24. Manufacturing Employee (2017) Disgraceful work environment. Pelham, NY, 21 Feb 2017. Available on Indeed https://www.indeed.com/cmp/Tiffany-&-Co./reviews/disgraceful-work-environment?id=ce3da2e98d65020d. Accessed 19 Aug 2020
25. Mead GH (1934) Mind, self and society. University of Chicago Press, Chicago
26. Model Maker (2019) Great co workers—a few exceptions, HR needs an overall and introduction to 2020. Pelham, NY, 5 Aug 2019. Available via Glassdoor https://www.glassdoor.com/Reviews/Employee-Review-Tiffany-and-Co-RVW30968744.htm. Accessed 2 July 2020
27. Palumbo S (2015) Precious cargo: Tiffany's supply chain security, 7 Dec 2015. Available via Loss Prevention Magazine https://losspreventionmedia.com/precious-cargo-tiffanys-supply-chain-network/. Accessed 11 Sept 2020
28. Polishing/Machine Engraving Specialist (2017) Tiffany use [*sic*] to be a fun place to work. Parsippany, NJ, 23 Sept 2017. Available via Indeed https://www.indeed.com/cmp/Tiffany-&-Co./reviews/tiffany-use-to-be-a-fun-place-to-work?id=879506e1a9d7ae60. Accessed 11 Aug 2020
29. Sennett R (2008) The craftsman. Yale University Press, New Haven and London
30. Sherman L (2019) Chanel surpasses $11 million in sales, dismisses rumours of imminent sale, 17 June 2019. Available via The Business of Fashion https://www.businessoffashion.com/articles/news-analysis/chanel-11-billion-sales-dismisses-rumours. Accessed 18 June
31. Solca L (2018) The truth about jewellery, 7 May 2018. Available via The Business of Fashion https://www.businessoffashion.com/articles/professional/the-truth-about-jewellery. Accessed 1 June
32. Thompson S (2006) Craft at a crossroads. Jewelry Arts & Lapidary Journal, Dec 2006, pp 30–36
33. Weaver M (2018) Medical students 'raised on screens lack skills for surgery', 30 Oct 2018. Available via The Guardian https://www.theguardian.com/society/2018/oct/30/medical-students-raised-on-screens-lack-skills-for-surgery. Accessed 18 June
34. Weisberg DE (2008) The engineering design revolution: the people, companies and computer systems that forever changed the practice of engineering. Available via CAD History https://www.cadhistory.net/. Accessed 7 July

Luxury and Sustainability: An Experimental Investigation Concerning the Diamond Industry

Matteo de Angelis, Cesare Amatulli, and Silvia Petralito

Abstract Over the years, the diamond industry has been marked by hotly debated scandals concerning the exploitative conditions afflicting the countries where diamonds were mined and extracted. Despite the establishment of formal certifications (e.g. the Kimberley Process), nowadays the diamond supply chain is still not completely devoid of ambiguities. Therefore, this research focuses on a new sustainable alternative: the lab-created diamond. The technology behind this man-made stone allows companies to obtain diamonds with the same physical, chemical and aesthetical properties of mined diamonds. In particular, we investigated how certain attributes specific to diamonds (namely perceived scarcity and authenticity) are perceived by consumers. We explored if the diamond typology presented (mined or lab-created) could affect respondents' perceived product authenticity by considering the potential moderating effect exerted by perceived product scarcity. Eventually, the diamond typology did not have any statistically significant effect on the level of perceived authenticity by itself while perceived scarcity played a crucial role in the relationship between mined diamonds and authenticity. When respondents sensed higher levels of perceived scarcity, mined diamonds were perceived as more authentic compared to lab-created diamonds. Therefore, we could affirm both the actual presence of a moderating effect exerted by scarcity and its pivotal involvement in decreeing mined diamonds as more authentic. Managers aiming at differencing these two typologies of diamonds could, therefore, emphasize perceived scarcity in their marketing and advertising campaigns.

Keywords Diamond industry · Sustainable luxury · Ethical diamonds · Lab-created diamonds · Mined diamonds · Perceived scarcity · Perceived authenticity

---

M. de Angelis · S. Petralito (✉)
Luiss University, Rome, Italy
e-mail: silvia.petralito@alumni.luiss.it

C. Amatulli
University of Bari, Bari, Italy

I. Coste-Maniere and M. Á. Gardetti (eds.), *Sustainable Luxury and Jewelry*,
Environmental Footprints and Eco-design of Products and Processes,
https://doi.org/10.1007/978-981-16-2454-4_9

1 Introduction

The following study is set within the framework of a new sustainable and ethical alternative in the diamond industry. Significant media coverage has been already dedicated to the exploitation involved in the mining process in Africa. The so-called "blood diamonds" have been traded to fund conflicts in Angola, the Democratic Republic of Congo and Liberia. As a consequence, the industry tried to address this alarming issue through the Kimberley Process [37] and the potential introduction of a Blockchain-based system to improve the diamonds' traceability [35]. Nevertheless, doubts regarding the ethical standards of the industry are still considerably present.

In the light of this concerning background, the diamond market is facing a new, more sustainable trend: the production of lab-grown diamonds. These diamonds are created inside highly controlled laboratories using advanced technological facilities and are able to perfectly duplicate the conditions through which diamonds naturally develop beneath the Earth's crust. These "synthetic" diamonds are made of actual carbon atoms, showing the peculiar diamonds' crystal structure and presenting identical chemical properties of natural diamonds. Even experts are unable to distinguish mined diamonds from lab-created ones without using high-tech equipment to detect their true origin.

The success of lab-grown diamonds will depend on persuading consumers, especially Millennials, and on how powerful mining companies will respond to the threat. In the next years, we will be able to establish whether diamond producers' marketing efforts effectively succeeded in creating a powerful concept as their 1948 slogan: "A Diamond is Forever". Miners, from De Beers to Alrosa, have the most to lose from any shift in the jewellery industry. In fact, different kinds of professionals involved in polishing, cutting and retailing activities can just possibly switch to lab-grown stones. However, according to Morgan Stanley's report regarding lab-grown diamonds [34], the diamond industry has spent less on marketing during the years. In the 1990s, the marketing budget of the industry's largest player corresponded to 5% of revenue while now the percentage has indeed decreased to 1%. To counteract this state of things, in 2015 seven of the world's largest miners formed the Diamond Producers Association and in 2016 launched the campaign "Real is Rare. Rare is Diamond".

Nevertheless, miners will probably need to refine their marketing investment in order to prevent a market share loss in favour of lab diamond producers. This rescheduling appears to be quite urgent, since diamonds have never really been a scarce commodity and the progressive customer awareness about it could change their perception. On the other hand, lab-grown diamond companies should also carefully consider their strategy. If they try to compete on cost, they could potentially destroy their own industry. Factories and labs compulsively producing high-quality diamonds on an everyday basis can clearly undermine their scarcity value. There is indeed a fine line between making consumers accept a lower price point and devaluing the overall market and profit margins.

In the light of the above, this study focuses on experimentally understanding the role that attributes like scarcity and authenticity have with respect to mined and lab-grown diamonds. The final findings can indeed allow diamond companies to better calibrate the importance of such product characteristics in their marketing and communication investments.

2 Sustainability Issues in Diamond Supply Chain

Even though companies involved in the mining, production and distribution of diamonds have always focused their communication on the emotional sphere connected to these stones, diamonds have also been associated with highly negative images. The so-called "conflict diamonds" or "blood diamonds" have surely represented the biggest scandal within the industry, thus increasing the general concern over sustainability and ethical matters. The above-mentioned denomination was diffused on a greater scale, thanks to the 2006 movie "Blood Diamond" directed by Edward Zwick and starring Leonardo di Caprio. The movie was nominated for five Academy Awards and managed to make the audience aware of the political and social unrest in African countries caused by the smuggling of diamonds. To give an idea of its impact, it is enough to consider that the World Diamond Council gave birth to a 15 million dollar campaign just a few months after the movie released in order to avoid consumer backlash (Carnagie Council, [12]).

As reported by Grynberg and Mbayi [22], rough diamonds have contributed to funding and financing violence since the 1950s. However, this dangerous phenomenon has been brought to public light approximately around the late 1980s and 1990s, especially for what concerns the diamond-funded armed violence in Angola and Sierra Leone. As described by Ceppi [13], a remarkable number of multinational diamond firms was involved in the purchase of gems from Angolan areas that were controlled by the União Nacional para a Independência Total de Angola (UNITA). By the mid-1990s, this phenomenon reached a dimension that was impossible to neglect. The UNITA was generating earnings of 700 million US$ per year from the trade of rough diamonds and, in Sierra Leone, the Revolutionary United Front (RUF) had an annual income of 125 million US$ obtained from the management of a remarkable portion of mining areas in the country.

Global Witness was the first organization to expose this state of things, back in 1998. It played a crucial role in the establishment of the Kimberley Process (hereafter KP), a government-led certification scheme put in place for cleaning the diamond trade. The scheme was officially launched in 2003, tightening up the controls over the stones' import and export. Nowadays, the KP consists of 54 participants representing 81 countries, with the European Union and its Member States counting as a single entity. However, the participation is also extended to organizations like the World Diamond Council and the Partnership Africa Canada.

In this perspective, the European Commission [17] certified that since 2003, the identifiable trade of conflict diamonds has declined from 15% to less than 1%. However, as Global Witness states on its website, the KP has shown severe limitations after its establishment. The first one to be mentioned is the narrow definition of conflict diamonds. The scheme indeed classifies them as "rough diamonds used by rebel movement to finance wars against legitimate governments". Therefore, the Kimberley Process is not entitled to protect human rights from other threats caused by the diamond trade. There has also been a clear denial to expand the field of application presented by the scheme itself, which is exclusively focused on rough stones. Consequently, when diamonds under scrutiny are successively cut and polished, they cannot be protected by the KP anymore. Global Witness even decided to resign from the role of official observer of the Kimberley Process in 2011, while defining the scheme as "an accomplice to diamond laundering". The Guardian then pointed out how the Kimberley Process "has failed on its own terms" since corruption and smuggling still represent an unpleasant reality [38].

Ultimately, Smillie [39] reported that even when facing episodes of "serious non-compliance" in Venezuela, Guinea, Lebanon and Zimbabwe, actions carried out by the Kimberley Process failed miserably with regards to the following issues: the total lack of compliance in Venezuela, where diamonds have been smuggled outside the country since 2006; the absence of internal controls in Guinea; significant inconsistency between the import and export statistics in Lebanon; smuggling and an overall shortage of lawful practices from exploration and licensing to minimal human rights in Zimbabwe.

3 Lab-Grown Diamonds

One of the most significant trends within the diamond industry is represented by the increasing magnitude of diamonds created within laboratories. In contrast to mined diamonds that are obtained through a process of geological evolution held within the Earth, lab diamonds are generated by the means of two possible different methods: *high-pressure, high-temperature* (HPHT) and *chemical vapour deposition* (CVD) [4].

In front of this new threat, the diamond industry should probably find a way to differentiate its offer. One of the alternatives suggested is that of positioning lab-grown diamonds more as a fashion than a luxury commodity. However, this scenario could result in a democratization of diamonds, especially by the means of a new CVD able to minimize the costs and maximize the quality. If in the past the production of a lab-grown diamond was about 4000$ per carat, now these expenses decreased to 300–500 $ [5].

In addition, two major events that occurred in 2018 should be mentioned. Firstly, the Federal Trade Commission provided the "Jewellery Guides" to avoid a potential deception exerted by diamond companies towards consumers. As highlighted by Danziger [14], this decision was supportive of scientific facts and not driven by the mining industry's interests, thus allowing consumers to have clear information before purchasing. The clarification provided by the FTC can be summarized as follows: a diamond is a diamond; it does not matter if the gem itself has originated from a process of natural extraction or from an "artificial" creation. Another significant transition is constituted by the removal of the "natural" attribute previously attached to diamonds. The old ruling was indeed stating "*A natural mineral consisting essentially of pure carbon crystalized in the isometric system*". FTC adopted this resolution since the two types of diamonds now have, thanks to more advanced technologies, "*the same optical, physical and chemical properties as mined diamonds*". Previously, companies involved in the production and selling of lab-grown diamonds were indeed forced to specify the stones' exact origin to consumers, by using attributes like "synthetic", "lab-created", or "lab-grown". The term "synthetic" has been specifically excluded since it is regarded as potentially misleading, implicitly suggesting a subtly negative connotation.

The second shift in the industry was represented by De Beers' launch of a lab-grown diamond brand, *Lightbox*, after an initial opposition towards this new phenomenon. This resolution was aimed at counteracting the new threat represented by lab-grown diamonds while contextually capitalizing on sales opportunities. In fact, as showed by Bain & Company's report [5], the sales of lab-grown diamonds only represent 2% of the total market, but the production will grow by 15–20% every year. Lightbox stones are available in three different colours (pink, blue and white) and their price point is currently 800$ per carat. This pricing choice is undoubtedly interesting since it is quite lower compared to the one offered by other lab-grown diamond companies.

According to Ali [2], the preference showed by consumers for either lab-grown or mined diamonds may change depending on their price and their perceptions regarding their scarcity and environmental impact. Moreover, one of the crucial elements that will affect the potential shift to lab diamonds is represented by the ability to develop a "signalling" effect on consumers [15]. Ali [3] also stated that the social signalling aspect, able to emphasize the relevance of sustainable consumption, should be analysed from two perspectives. Firstly, the act of purchasing diamonds has traditionally been associated with gift-giving rather than with an individual buying process. Therefore, the author suggests how the communicative relationship between the gift-giver and the gift recipient is considerably important. Secondly, the act of signalling value is then enhanced and encouraged during social interactions when people usually tend to ask questions regarding the origin and worth of these items. Eventually, it is suggested that just like cultured pearls, lab-created diamonds could witness a similar trend by becoming fashionable.

4 Marketing Campaigns in the Diamond Industry

Luxury has always aimed at being and appearing timeless [28]. Timelessness indeed represents one of the main distinguishing factors between luxury and fashion consumption, since the eternal dimension within which luxury goods are embedded allows consumers to discern luxury itself from what it is perceived as ephemeral

[26]. When considering diamonds, even if these stones can theoretically be cut, shattered or deprived of their colour, the advertising campaign behind the "A Diamond is Forever" slogan invested this gem with an eternal aura.

In the 1930s, De Beers and its agency N. W. Ayer managed to psychologically influence American consumers by implicitly leading them to the purchase of diamonds as engagement rings. The clarity, size and quality of the stone were then used as a benchmark for assessing the love shared by the couple. The main objective was that of reflecting the timelessness attribute over the romantic relationship itself and portraying the image of rock-solid love [16]. As noted by Bergenstock [7], the marketing plan carefully orchestrated by De Beers managed to convey a consistent message to the public, made of wealth, romance, elegance and eternity. This concept has been recognized as extremely brilliant by the special edition of Advertising Age, "The Advertising Century", which ranked De Beers' campaign at the sixth place out of 100.

Nevertheless, the perception of perpetuity has been always associated with *mined* diamonds, which need from 1 to 3.3 billion years to form under the Earth's crust. Clearly, the same cannot be said with regards to lab-grown diamonds, which only need a couple of weeks to be produced. The timelessness attribute was also reinforced by De Beers' brand name: Forevermark, which was launched in 2008. Therefore, to counteract the threat posed by lab-grown diamonds, in 2016 the Diamond Producers Association presented a new advertising campaign bearing the slogan "Real is Rare. Rare is a Diamond". This strategic choice was also aimed at grabbing the attention of the Millennial generation despite the scandals that characterized the industry.

During the first 12 months, the Diamond Producers Association's paid media campaign generated 1.5 billion media impressions among the desired target, belonging to the age range of 21–34 years old. Television, digital and social media have been the focus of a $40 million worth investment. A smaller percentage, approximately 10%, has been dedicated to print advertising and a 5% to strategic out-of-home placements such as cinema. As a result, the Diamond Producers Association decided to increase the global budget from 40 to 70 US$ million, with a major focus in the United States [40]. The pivotal goal of the new marketing attempt was indeed to portray a more relevant scenario to the Millennial customer. Different kinds of couples are therefore depicted in the campaign commercials in order to convey a sense of inclusivity together with alternative types of commitment, love and connection [8]. This campaign also emphasized different themes that are linked to the evolution of our society, such as the concept of self-gifting. The spots named "For Me, From Me" launched in 2019 celebrate the self-purchase process that today's women carry out at all life stages. Reportedly, the self-purchase carried out by women represents one-third of the entire diamond jewellery sales within the United States (Diamond Producers Association).

5 Scarcity Effects and Their Importance on Diamond Perception

The advertising campaign "Real is Rare" is based on the application of the evergreen commodity theory first mastered by Brock [11] who affirmed that "*any commodity will be valued to the extent that it is unavailable*". People's preference for unavailable products was then further analysed by Fromkin and Snyder [19] who suggest that individuals have a generalized need of feeling unique and are willing to make themselves distinct from others. Consumer behaviour indeed should be investigated by understanding the meanings that consumers attach to material possessions. Already back in 1980 (p. 472), Tuan stated: "*Our fragile sense of self needs support and this we get by having or possessing things because, to a large degree, we are what we have and possess*". Since our possessions are a major contributor and reflection of our identities [6], this means that a scarcely available product could be perceived as a metaphorical representation of the individual uniqueness. As shown by the meta-analysis conducted by Lynn [32], scarcity emphasizes the value of anything that can be possibly owned. Marketers can consequently enhance the perceived value of products, services and promotions in general by psychologically manipulating the notions of rarity associated with them. In particular, the above-mentioned research found that the enhancement of value carried out by scarcity is considerably greater when respondents' need for uniqueness is higher. Therefore, as the "Real is Rare" campaign demonstrates, the concept of such communication is mainly directed to the target of people feeling a greater need for uniqueness. However, Kapferer [29] managed to illustrate which factors could contribute the most to the dimension of the desirability of a luxury brand even when rarity becomes a less salient attribute for the brand itself. The findings suggest that, in modern luxury, a product's emotional charge actually represents a cornerstone for brand building. At the same time, the author shows the crucial importance of captivating customers by attributing the brand with a high seductive symbolic capital. Indeed, as pointed out by Kapferer [27], even the retail expansion strategies which characterized luxury brands in the last years could represent good news, if not for the basic equation according to which luxury equals rarity. According to this view, a greater penetration of the product within the market will at first enhance the brand visibility and then cause the luxury status dilution after reaching a break-even point.

Moreover, Mittone and Savadori [33] demonstrated that the so-called scarcity bias occurs when the subjective value of a good increases simply because of its unavailability. In the study, scarcity has been theorized as the presence of limited resources and competition on the demand side.

The results showed that the rare good was marked by a higher willingness to accept the price by the respondents who chose it compared to those that opted for the abundant good.

With the campaign "Real is Rare", the natural diamond industry is implicitly trying to position its good as the authentic luxury alternative in consumers' minds compared to the "artificial" man-made diamonds. De Beers' chief executive Bruce

Clevear has no doubts concerning this issue: "Synthetics are not rare or given at life's great moments. Nor should they be" [36]. Eventually, scarcity represents a defining characteristic for a luxury product [25] and it also serves as the very cornerstone of luxury gems and jewellery industry in general [1]. As Kapferer [26] emphasized, the value of a luxury brand is intrinsically linked to rarity and price, since these elements help distinguishing buyers from non-buyers.

6 Perceived Authenticity

Another issue that needs to be addressed is represented by the perceived authenticity of lab-grown diamonds in the eyes of consumers. Even though mined and lab diamonds present the same chemical and physical characteristics, the fact itself of being man-made could create a disadvantage for the second ones. However, the Federal Trade Commission considerably helped lab diamond producers by forbidding the use of adjectives like "synthetic" when referring to them. Although it may seem just a descriptive clarification, this change constitutes a crucial shift since it prevents confusion on behalf of consumers. Nonetheless, it should be analysed if consumers nowadays really perceive these two diamonds as equally authentic in their nature.

Authenticity, as a concept, has been at the centre of several studies aimed at investigating consumer behaviour. Pine and Gilmore [20] dedicated a whole book to this topic, considered by them as a business imperative. The authors recognize how people today tend to immediately categorize individuals, products or experiences as real or fake. This tendency, however, underlines a common understanding: what is perceived as real—and therefore authentic—is valued. Beverland et al. [9] provide a thorough consideration of the different challenges that the quest for authenticity poses to brand managers. According to the authors, in order to convey an authentic imagery, managers should locate the brands within communities and sub-cultures while appealing to timeless values.

However, how can we define authenticity? Trilling [41] understood the authenticity of a good in terms of its perceived genuineness and positive valuation. Firat et al. [18] conceptualized the need for authenticity as a natural and inevitable response to the threating inauthenticity specific to postmodernism. As Liao et al. [30] pointed out, today consumers put emphasis on authenticity as an attribute since mass production could possibly impair the product value. In this perspective, we can suppose that luxury buyers interested in the purchase of diamonds could undervalue lab-made diamonds' authenticity because of the "industrial" nature characterizing their production.

The distinction that Grayson and Martinec [21] made regarding the matter of authentic marketing offerings is also quite pertinent. The authors identify two different kinds of authenticity: the first one takes place when an item presents a factual and spatiotemporal connection to history—intended as a context in which people and items have an interaction: *identical authenticity*; on the contrary, when a product represents a reproduction of the original and, therefore, accurately looks like the original on a physical perspective, the so-called *iconic authenticity* takes place. We can imagine how crucial the role of authenticity is within the context of a luxury purchase, especially when confronting the "original" luxury product with its sustainable alternative. The design of a luxury item thus also heavily contributes to the creation of its authentic perception, by reinforcing its consistency and individuality [24].

In the light of what has been already mentioned, some considerations can be made. As noted above, the strategic aim carried out by the "Real is Rare" campaign is represented by the association between the attributes of *scarcity* and *authenticity*. This study wants to investigate the relationship between these two variables and the two possible alternatives, namely lab diamonds and mined diamonds. Mining companies are surely trying to emphasize the authenticity of mined diamonds—compared to the ones created in a lab—by stressing their rarity. However, mined diamonds' scarcity is here addressed in purely subjective terms.

This clarification is needed since the communication campaigns carried out by the diamond industry in the last decades have made consumers perceive diamonds as extremely scarce, regardless of their actual availability in the mines. The number of diamonds offered in the market has been artificially controlled for years, and this strategic move heavily contributed to reinforcing the perception of rarity already conveyed through the media. It is also worth noticing that lab-grown diamonds represent a quite new commercial alternative if compared to the mined counterparts. Therefore, the strong associations developed during the years could be inevitably hard to untie from people's minds in a relatively short time span. Eventually, in the light of the previous literature review and the above-mentioned considerations, it is possible to develop the following hypotheses:

H1: Higher (lower) levels of perceived product scarcity determine an increase (decrease) of perceived product authenticity for mined (lab-grown) diamonds.

H2: Perceived product scarcity moderates the relationship between the type of diamond (mined vs. lab-grown) and perceived product authenticity. When mined diamonds are perceived as scarcer, they are also perceived as more authentic products.

7 Experiment

The following empirical analysis presents a series of specific goals. Firstly, we will investigate if the perceived authenticity is higher for mined diamonds when compared to lab-created ones. Secondly, we will attempt to determine whether there is a reliable connection between perceived scarcity and perceived authenticity. The focus will be on certifying if higher levels of perceived scarcity caused higher levels of perceived authenticity. Specifically, the last observation was developed with regards to mined diamonds, because of the overall higher levels of scarcity associated with these stones and the ways through which they are extracted and therefore commercialized. The

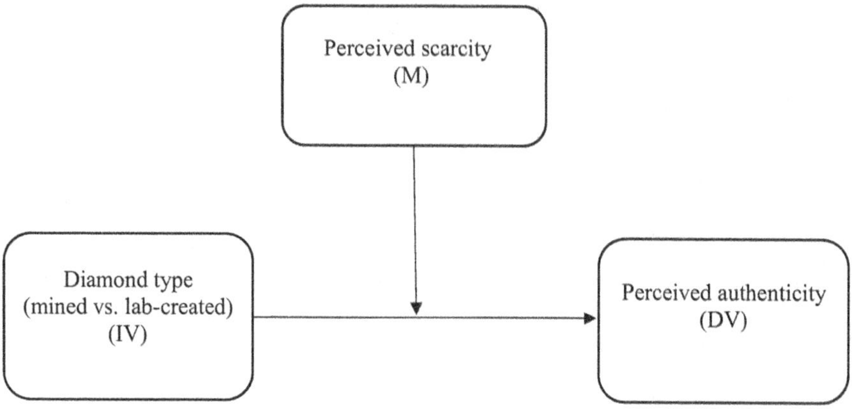

Fig. 1 Moderation model with perceived scarcity as moderator and perceived authenticity as dependent variables

pivotal intention of this analysis is indeed the determination of the moderating effect of perceived scarcity between mined diamonds and perceived authenticity.

A moderation model has been employed to analyse the data provided and test the hypotheses that have been formulated above. The model reflects a causal sequence in which X (diamond type) directly affects Y (perceived authenticity) by interacting with moderator M (perceived scarcity). See Fig. 1.

8 Procedure and Measures

The research consisted in an online experiment in which respondents have been randomly assigned to different experimental conditions and asked to complete a questionnaire. The survey has been created and distributed to participants through Qualtrics while data were then analysed via SPSS. Two hundred and four participants (105 females and 99 males) have provided complete responses to the survey. From a socio-demographic point of view, eighty-eight respondents were aged between 18 and 24, one hundred and twelve were aged between 25 and 34, and only four of the participants were aged between 35 and 44. As regards the monthly personal income, it has been recorded that sixty-one respondents indicated a monthly income lower than 1000 €; one hundred and one of the surveyed individuals indicated a monthly income between 1000 and 1999 €; thirty-four between 2000 and 2999 € and eight of them showed a higher monthly income (3000–3999 €). One hundred and seven respondents were randomly assigned to the lab-created diamond condition and the remaining ninety-seven to the mined diamond condition. Before reading the scenario, however, participants supplied information regarding their Corporate Social Responsibility (hereafter CSR) expectations of luxury brands. The necessity to investigate internal validity is related to the potential presence of confounding variables within the study itself. In this perspective, a higher level of this variable could influence individuals and make them perceive mined diamonds more negatively and lab diamonds from a more positive standpoint since they could be aware of the ethical concerns related to diamonds' extraction. Participants' CSR expectations have been assessed through a scale developed by Bhattacharya and Sen [10] consisting of seven items, namely "Luxury brands have to protect endangered species", "Luxury brands must take part in the efforts made to reduce water consumption", "It is important for luxury brands to use recycled packaging", "It is important for luxury brands not to exploit their employees", "Luxury brands have to fight against climate change (production and transportation of products), "Luxury brands must take part in the efforts made to reduce energy consumption" and "It is important for luxury brands to educate their employees about their impact on the environment" (1 = "strongly disagree", 7 = "strongly agree", $\alpha = 0.95$). Following this stage, respondents were presented with a brief and neutral introduction regarding the subsequent steps of the survey. Individuals were then randomly assigned to the two different experimental conditions. In each case, respondents read a scenario about a luxury brand involved in diamonds' production. Specifically, the independent variable was manipulated by communicating to participants that a new luxury brand named "Brillance" had recently launched in the market a diamond ring with either a mined or lab-created diamond. Respondents have contextually been presented with images of the diamond ring randomly assigned to them. There was no risk of creating an unintended aesthetic preference towards one ring or the other since the two typologies of diamonds cannot be distinguished by the naked eye. The chosen images were indeed showing the same ring.

After reading the scenarios, participants responded to the questions related to the product they have just seen, assessing perceived authenticity and scarcity. As mentioned above, perceived authenticity represents the dependent variable of this study. It has been measured through the adaptation of a scale developed by Lubberts [31] since the ones that have been previously developed by other authors mainly focused on the brand's perceived authenticity. In the context of this research, we indeed examined the perceived authenticity attributed to the product itself, rather than the brand. The scale items were extremely coherent with the aim of the research and the scale itself presented a remarkable internal consistency ($\alpha = 0.91$) which further supported this choice. Specifically, the 7-point Likert scale items are the following (1 = strongly disagree, 7 = strongly agree): "Quality is central to this product", "Only the finest ingredients/materials are used in the manufacture of this product", "It feels like craftsman skills have been retained in the production of this product", "The firm Brillance is committed to retaining long-held quality standards for this product", "The product makes the impression of being natural", "The product makes a genuine impression" and "The product does not seem artificial". On the other hand, scarcity has been measured through a 7-point Likert scale (1="totally disagree", 7="totally agree", $\alpha = 0.60$) provided by Janssen et al. [25], composed by three items: "This product is unique, original", "This product is made of rare and precious materials" and "This product can easily be found". As a moderator variable, we expect perceived scarcity to alter the strength of the causal relationship between X and Y. It goes without saying that the product type constitutes a categorical variable while both perceived scarcity and authenticity represent metric variables. As already mentioned at the beginning of this paragraph, respondents were also asked to provide some socio-demographic information, specifically regarding their income, gender and age. In this case, the net monthly personal income of the individual has been taken into account, by providing 6 possible income ranges: (<1000$, 1000–1999$, 2000–2999$, 3000–3999$, 4000–4999$, 5000 and more). On the other hand, when assessing respondents' age, the following age ranges have been included for the analysis: 18–24, 25–34, 35–44, 45 and more.

9 Results

Before proceeding with the moderation analysis, a series of correlation analyses have been carried out to better examine the correlation existing between the items of the scales. When considering the control variable, all items were reciprocally, significantly and positively correlated to each other. The correlations ranged from a minimum of 0.400 to a maximum of 0.791. Therefore, a factorial analysis was conducted to reduce the items to a single factor, which has been successively denominated as CSR. The same kind of correlation assessment has been executed with the moderator: perceived scarcity. Notably, the three items were reciprocally and significantly correlated to each other, showing both positive and negative signs. The correlations, in absolute value, tended to vary from a minimum of −0.609 to a maximum of −0.841. As similarly mentioned above, a confirmatory factor analysis was conducted to reduce all these items to a single factor that will be called "Scarcity". The same process was repeated with the items composing perceived authenticity. Even in this scenario, all the items were reciprocally, significantly and positively correlated to each other. The correlations varied from a minimum of 0.535 to a maximum of 0.904. Once again, a confirmatory factor analysis will be helpful for narrowing these items to one factor that will be referred to as "Authenticity".

As already mentioned, by the means of a confirmatory factor analysis, the reported items were then reduced to factors. In particular, we expected to extract a single factor for each group of items (one factor for CSR, one for scarcity and eventually one for authenticity). The first factor to be obtained was the CSR, by using a principal component analysis. In this case, no rotations were needed since a single factor will be extracted in the end.

The Kaiser-Meyer-Olkin (hereafter KMO) test, in particular, represents the relationship between the square of the correlation between the variables and the square of the partial correlation between the variables; in this case, a value which is equal to 0,910 is optimal since the KMO score ranges from 0 to 1, and the closer to 1, the better. Bartlett's test instead has as a hypothesis the assumption of sphericity which has to be refused exactly like in this case ($p < 0.001$).

Communalities represent the percentage of variability of the original items that is explained by the factor analysis. The values appear to be all quite satisfying since they are all greater than 65% except for the item "It is important for luxury brands not to exploit their employees" that has a communality of 42,6%. We then analysed how much of the total variance was explained by the extracted factor. In this case, the factor explained almost the 69% of the original variance of the 7 items. As a consequence, the result appears to be satisfactory since the remaining unexplained variance represents only 31%. Moreover, all items resulted to be positively correlated to the extracted factor. The lowest loading was represented by the item "It is important for luxury brands not to exploit their employees" that was indeed showing the lowest communality as well. Before deciding whether to keep this item or not, Cronbach's Alpha analysis has been conducted in order to determine and assess the reliability of the scale composed by these items. We noticed how Cronbach's Alpha value was indeed high (0.922). In addition, when removing the item "It is important for luxury brands not to exploit their employees", a minor increase of Cronbach's Alpha was detected. Therefore, all items were maintained while saving the scores of the CSR factor, which will be employed in the moderation analysis. Successively, a second factor analysis has been conducted for Scarcity, following the previous process. In this case, the KMO showed an acceptable value (0.671). Again, the Bartlett test's hypothesis requires that the assumption of sphericity needs to be rejected exactly like our results suggest (p-value < 0.001). The communalities, representing the percentage of original items' variance explained by the factor analysis, showed satisfactory outcomes (all higher than 70%). We determined how much of the total variance is explained by the extracted factor. In this case, the factor explains almost 82% of the three items' original variance. Therefore, the findings were very encouraging since the remaining unexplained variance attested to 18%. The first and the third items of the Scarcity scale resulted to be positively correlated to the extracted factor, as well as expected. On the other hand, the second item appeared to be negatively correlated to the extracted factor in a coherent way, since it has an inverse polarization compared to the other items. Moreover, all three loadings display high features in absolute value. Considering these observations, the second item of the scale has been inverted and, successively, Cronbach's Alpha analysis has been carried out in order to confirm the reliability of the scale composed by these items. When looking at Scarcity items, it is possible to notice how Cronbach's Alpha value is also remarkable (0.886). By removing the item "This product is unique, original", we noted a negligible increase of Cronbach's Alpha itself. As a consequence, all the items were kept, and the Scarcity factor score was saved in order to be further used in the context of the moderation analysis.

The final factor analysis has been conducted for the Authenticity factor. In this instance, the KMO displayed a value of 0.895 which is surely high and optimal since 1 is the maximum value that it could possibly reach. The Bartlett test, again, allows rejecting the assumption of sphericity since the p-value < 0.001. Moreover, even in this case, the communalities present satisfying results since the values are all greater than 65%. We then investigated the total variance explained by the extracted factor. This time, the factor explained approximately 78% of the original variance of the 7 items. The outcome appeared to be highly positive since the remaining unexplained variance was only about 22%. All the items were positively correlated to the extracted factor, exactly as expected. Once again, Cronbach's Alpha analysis was executed in order to assess the reliability of the scale constituted by these items. When it comes to the Authenticity's items, Cronbach's Alpha value reaches a high feature (0.946). When removing the item "Only the finest ingredients/materials are used in the manufacture of this product", a negligible increase of Cronbach's Alpha was found. Therefore, all items were kept, and the Authenticity factor score was saved so that it will be successively employed in the moderation analysis.

In order to conduct the moderation analysis, we used the PROCESS SPSS Macro developed by Hayes ([23], Model 1). First of all, it is worth mentioning that the model was actually able to explain a remarkable percentage of variance, even more than 70% (R-squared = 0.72). Results show that the direct effect exerted by the independent variable on the dependent one was not statistically significant ($p = 0.33$). This finding therefore excludes the presence of a significant direct effect exerted by the diamond typology on perceived authenticity. In addition, the role of the control variable representing respondents' CSR expectations of luxury brands was not statistically significant ($p = 0.14$). Therefore, we can assume that participants' personal convictions regarding the level of sustainability efforts carried out by luxury brands did not interfere with the other variables involved in the study.

Importantly, the effect of the moderator on the dependent variable resulted to be statistically significant and positive ($b = 0.97$; $p < 0.001$), and the effect of the interaction (between the independent variable and the moderator) had a significant and negative effect ($b = -0.43$; $p = 0.03$) on perceived authenticity. Even the test to verify whether the increase of R-squared due to the interaction confirmed that the interaction inclusion contributed to a statistically significant betterment of the R-squared. In order to better understand the nature of the moderation itself with respect to our formulated hypotheses, a further analysis was conducted to assess the effects of the independent variable on the dependent variable at different levels of the moderating variable.

When the respondent presents a low score on the Scarcity factor, the coefficient of the independent variable is equal to 0.24, it is not statistically significant ($p = 0.45$) and the estimated Authenticity is very low, equal to −1.07 for the mined diamond and −0.83 for the lab-created diamond. When the respondent has an average score on the Scarcity factor, the coefficient of the independent variable is negative and has a value of −0.91, it is not statistically significant ($p = 0.33$) and the estimated Authenticity is slightly negative, equal to −0.09 for the Mined diamond and −0.29 for the Lab-created diamond. When the respondent shows a high score on the Scarcity factor, the coefficient of the independent variable is negative and equal to −0.62, it is statistically significant ($p = 0.007$) and the estimated authenticity is positive: 0.87 for the mined diamond and 0.25 for the lab-created diamond.

As a consequence, we found out that there is actually a strong moderation effect exerted by Scarcity and that the impact of the product is extremely related to the level of scarcity perceived by the respondent. Therefore, results confirm both H1 and H2. When mined diamonds are perceived as scarcer, they are also perceived as more

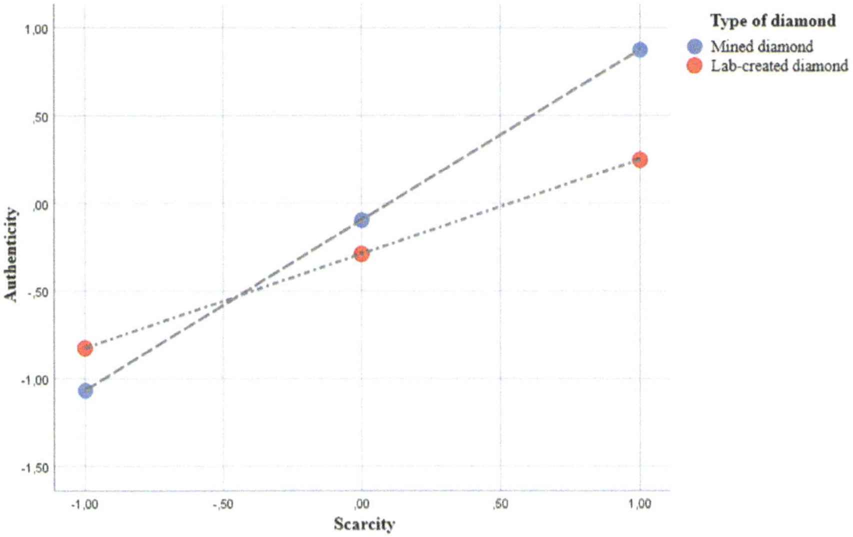

Fig. 2 Perceived authenticity on the basis of scarcity and type of diamond

authentic products. The obtained results are graphically represented in the following Fig. 2.

10 General Discussion

This research investigated consumers' perception of authenticity towards two different kinds of jewellery pieces: one featuring a less sustainable alternative (mined diamond) and another showing its sustainable and ethical counterpart (lab-created diamond). Specifically, we explored whether the diamond typology presented could affect customers' perceived product authenticity by considering the moderation effect exerted by the perceived product scarcity. This investigation wanted to include in the analysis customers' corporate social responsibility of luxury brands to assess whether this element could affect in any way the above-mentioned variables.

We started from the assumption that—because of years of continuous and pervasive communication and marketing strategies—mined diamonds were associated with a higher degree of both authenticity and scarcity compared to the ethical and way more recent alternative represented by lab diamonds. For this reason, one of the first goals of this study was that of assessing whether the condition showing the mined diamond ring was causing by itself a higher level of perceived product authenticity.

By looking at the results, at least within this experimental context, the type of diamond with which participants were presented did not have any kind of statistically significant effect on the level of perceived authenticity by itself. This result is also quite understandable if we think that the sustainable alternative presents exactly the same aesthetic connotations as the extracted stones. One of the most interesting results, however, is related to the crucial role that perceived scarcity played in the relationship between mined diamonds and authenticity. Indeed, when respondents perceived higher levels of perceived scarcity, mined diamonds were perceived as more authentic compared to lab-created diamonds. This observation allows us to affirm not only the actual presence of a moderating effect exerted by scarcity, but also its pivotal involvement in decreeing mined diamonds as more authentic.

11 Managerial Implications

From a managerial standpoint, this research provides luxury companies operating in the diamond or jewellery industry with interesting implications. In fact, the launch of a new sustainable alternative in the market could represent a double-edged sword for any luxury brand. This kind of concern can assume a considerable entity, especially when examining diamonds, as well as any product composed of a highly precious metal that can potentially leave space to doubts when analysing the ethical policies behind its extraction.

In the light of what has been previously said regarding the communicative strategies employed in the diamond industry, the results could suggest which attributes should be emphasized when communicating to potential consumers. Chiefly, this investigation denotes the pivotal role played by product scarcity perception on the way consumers consider the overall product authenticity. In fact, higher levels of perceived scarcity determined an increased level of perceived product authenticity for mined diamonds. Consequently, diamond companies involved in the extraction and sale of these precious stones should highlight the scarcity attribute when presenting advertisements or other marketing efforts. In this perspective, the Diamond Producers Association has already paved the way. It has been already remarked how this organization responded to the potential threat of lab-created diamonds through a campaign focused on mined diamonds' rarity compared to their ethical counterpart: "Real is Rare". Nonetheless, the relevance of scarcity pointed out by this study could in a way reverse the slogan in "Rare is Real". Scarcity should therefore be conceptualized by managers as a crucial point of difference between mined diamonds from lab-created diamonds.

This consideration is extremely important when also taking into account that the two product typologies look the same to the human eye and, therefore, an eventual differentiation on an aesthetic level would not be possible for managers to implement. Another readable outcome is represented by the lack of a potential confounding effect of the CSR expectations of luxury brands. In this regard, managers could focus more on conveying to customers the importance of the other attributes without excessively being concerned about customers' expectations with respect to ethical and social matters revolving around the company itself.

12 Limitations and Future Research Directions

This study, however, presents some limitations that could offer further paths for future research. When considering the conceptualization of the model we employed, it is possible to notice the selection of a single moderator. Future experimental analyses could surely examine how different and additional factors might leverage the impact exerted by product type on perceived authenticity. In this view, not only future works can develop other moderation models by replacing the scarcity perception with another attribute, but they could also investigate the interplay between scarcity and another variable.

Moreover, this analysis does not look at the price difference currently existing between the two types of diamonds on the market. Lab-created diamonds are indeed considerably low-priced compared to the extracted ones, as previously pointed out. Perhaps, informing respondents about this additional factor could sensibly affect their perceptions regarding the other variables involved. Future projects could therefore consider the role that customers' willingness to pay may have with respect to different diamonds, while always bearing in mind their essential equality in terms of physical, chemical and aesthetic connotations.

It must also be considered that the questionnaire has been distributed to European respondents only, most of which were Italian. Similar research could be carried out by distributing it in other regions to detect potential cross-cultural differences and perhaps in the U.S. where the diamond market finds its most affectionate and passionate customers. The sample examined in this investigation is also very homogenous, both in terms of age and income ranges; the access to a more heterogeneous population could surely provide other interesting results.

Eventually, since the questionnaire has been distributed exclusively online, future works could test the result of this dissertation by replicating the study in the field with real luxury consumers and by showing real diamonds instead of pictures.

13 Conclusions

At first, this research focused on analysing the main sustainability issues related to the diamond industry. If it is true that these stones have been associated during the years with highly emotional and romantic attributes, it is also undisputable how they have also been tied to several scandals. The "blood diamonds" phenomenon has been brought to public knowledge mostly in the last decade, thus shedding light on a general situation of exploitation in the countries where the diamonds were mined and extracted.

The poor condition in which individuals were forced to work was not—however—the only issue which got mining companies a negative reputation. The generally uncontrolled supply chain of this industry allowed diamond trade to fund civil wars and arm violence since the 1950s, especially in poor realities like Angola and Sierra

Leone. As mentioned above, several initiatives and certifications have been carried out in order to mitigate the entity of such dramatic scenarios, such as the Kimberley Process. Despite the efforts, diamonds' smuggling still represented an unpleasant reality in many countries, even though these certifications significantly contributed to making it more difficult.

However, besides the implementation of formal regulations, a different kind of sustainable alternative has been proposed. In fact, a new diamond typology has started to make its way into the market: the lab-created diamond. As it is possible to assume from the name itself, this particular diamond type is created within the environment of a laboratory by taking advantage of specialized machines able to replicate the conditions in which "natural" diamonds form below the Earth's crust. This technology allows companies to obtain diamonds showing exactly the same physical and chemical properties as mined ones, without exploiting any workforce or funding any armed conflict, at a lower price point as well. This more sustainable substitute could have the potential to disrupt the market, by putting a threat over those companies that are involved in the mining and sale processes.

This research, in particular, wanted to analyse the role that specific diamonds' attributes had in the eyes of the consumers, with a focus on scarcity and authenticity. More in depth, the submitted survey randomly presented participants with a mined or lab-created diamond, asking targeted questions to investigate the role of their perceived scarcity on their authenticity. After examining the obtained data, some interesting conclusions were indeed drawn. In the context of this study, perceived scarcity had a strong moderating role between mined diamonds and their authenticity. In particular, the scarcer mined diamonds were perceived, the higher was their perceived authenticity. This observation could result to be quite useful for mining companies that are trying to keep their market share intact with respect to the threat imposed by lab diamonds. Miners could emphasize the scarcity attribute in their communication, thus conveying this rare connotation as an added and distinguishing value with respect to lab diamonds. However, this field is still open to future research and investigations that could consider other product attributes, able to provide managers with a more complete picture of the forthcoming trends and opportunities.

References

1. Ali SH (2010) Treasures of the earth: need, greed, and a sustainable future. Yale University Press
2. Ali SH (2011) Ecological comparison of synthetic versus mined diamonds. Working paper, Institute for Environmental Diplomacy and Security, University of Vermont
3. Ali SH (2017) The ecology of diamond sourcing: From mined to synthetic gems as a sustainable transition. J Bioecon 19(1):115–126
4. Bain & Company (2011) The global diamond industry. https://www.bain.com/contentassets/dab98eb7f88c4bbf8b9c3e48b816c362/pr_bain_report_the_global_diamond_industry.pdf. Accessed July 2019

5. Bain & Company (2018) The global diamond industry 2018. https://www.bain.com/contentassets/a53a9fa8bf5247a3b7bb0b10561510c2/bain_diamond_report_2018.pdf. Accessed July 2019
6. Belk RW (1988) Possessions and the extended self. J Consum Res 15(2):139–168
7. Bergenstock DJ, Maskulka JM (2001) The De Beers story: are diamonds forever? Bus Horiz 44(3):37–44
8. Bergstein R (2016) What the diamond industry's new campaign, "Real is Rare", says about marketing luxury to Millennials. Forbes.com https://www.forbes.com/sites/rachellebergstein/2016/10/19/what-real-is-rare-the-diamond-industrys-new-campaign-says-about-marketing-luxury-to-millennials/#427d374e2107. Accessed Aug 2019
9. Beverland MB, Farrelly FJ (2010) The quest for authenticity in consumption: consumers' purposive choice of authentic cues to shape experienced outcomes. J Consum Res 36(5):838–856
10. Bhattacharya CB, Sen S (2004) Doing better at doing good: when, why, and how consumers respond to corporate social initiatives. Calif Manag Rev 47(1):9–24
11. Brock TC (1968) Implications of commodity theory for value change. In: Psychological foundations of attitudes. Academic Press, pp 243–275
12. Carnagie Council for Ethics in International Affairs (2008) Ethics on film: discussion of "Blood Diamond". https://www.carnegiecouncil.org/publications/ethics_onfilm/0002. Accessed Aug 2019
13. Ceppi JP (2000) Conducting business in war-prone areas. War, money and survival, pp 40–43
14. Danziger PN (2018) What The FTC Diamond ruling means for consumers and diamond marketers. Forbes.com. https://www.forbes.com/sites/pamdanziger/2018/08/02/what-the-ftc-diamond-ruling-means-for-consumers-and-diamond-marketers/#1dfa36e20fa9. Accessed Sept 2019
15. Delgado MS, Harriger JL, Khanna N (2015) The value of environmental status signalling. Ecol Econ 111:1–11
16. Epstein EJ (1982) Have you ever tried to sell a diamond? Atl Mon 23:363
17. European Commission (2018) Eu chairmanship of the Kimberley Process 2018, Supporting peace and development through the sustainable use of natural resources. https://ec.europa.eu/fpi/sites/fpi/files/factsheet-kimberleyprocess-rev.pdf. Accessed Oct 2019
18. Firat AF, Venkatesh A (1995) Liberatory postmodernism and the reenchantment of consumption. J Consum Res 22(3):239–267
19. Fromkin HL, Snyder CR (1980) The search for uniqueness and valuation of scarcity. Social exchange. Springer, Boston, MA, pp 57–75
20. Gilmore JH, Pine BJ (2007) Authenticity: what consumers really want. Harvard Business Press
21. Grayson K, Martinec R (2004) Consumer perceptions of iconicity and indexicality and their influence on assessments of authentic market offerings. J Consum Res 31(2):296–312
22. Grynberg R, Mbayi L (eds) (2015) The global diamond industry: economics and development, vol 2. Springer
23. Hayes AF (2017) Introduction to mediation, moderation, and conditional process analysis: a regression-based approach. Guilford Publications
24. Hitzler PA, Müller-Stewens G (2017) The strategic role of authenticity in the luxury business. Sustainable management of luxury. Springer, Singapore, pp 29–60
25. Janssen C, Vanhamme J, Lindgreen A, Lefebvre C (2014) The Catch-22 of responsible luxury: effects of luxury product characteristics on consumers' perception of fit with corporate social responsibility. J Bus Ethics 119(1):45–57
26. Kapferer JN (1997) Managing luxury brands. J Brand Manag 4(4):251–259
27. Kapferer JN (2012) Abundant rarity: the key to luxury growth. Bus Horiz 55(5):453–462
28. Kapferer JN, Bastien V (2009) The specificity of luxury management: Turning marketing upside down. J Brand Manag 16(5–6):311–322
29. Kapferer JN, Valette-Florence P (2016) Beyond rarity: the paths of luxury desire. How luxury brands grow yet remain desirable. J Prod Brand Manag 25(2):120–133

30. Liao S, Ma YY (2009) Conceptualizing consumer need for product authenticity. Int J Bus Inf 4(1)
31. Lubberts T (2018) The impact of product cue manipulation and consumer product involvement on product authenticity evaluations. Dissertation, Wageningen University
32. Lynn M (1991) Scarcity effects on value: a quantitative review of the commodity theory literature. Psychol Mark 8(1):43–57
33. Mittone L, Savadori L (2009) The scarcity bias. Appl Psychol 58(3):453–468
34. Morgan Stanley (2016) Global diamonds: game of stones—lab vs. pipe. https://www.morganstanley.com/ideas/diamond-market-lab-grown-disruption. Accessed Aug 2019
35. Nofer M, Gomber P, Hinz O, Schiereck D (2017) Blockchain. Bus Inf Syst Eng 59(3):183–187
36. Paton E, Friedman V (2018) "Diamonds Are Forever" and made by machine. The New York Times.com https://www.nytimes.com/2018/05/29/business/de-beers-synthetic-diamonds.html. Accessed Sept 2019
37. Price TM (2003) The Kimberley Process: conflict diamonds, WTO obligations, and the universality debate. Minn J Global Trade 12:1
38. Rhode D (2014) The Kimberley Process is a 'perfect cover story' for blood diamonds. The Guardian.com https://www.theguardian.com/sustainable-business/diamonds-blood-kimberley-process-mines-ethical. Accessed Sept 2019
39. Smillie I (2011) Assessment of the Kimberley Process in enhancing formalization and certification in the diamond industry-problems and opportunities. German Federal Ministry for Economic Co-operation and Development, Berlin
40. The Centurion (2018) One year later, DPA's successful "Real is Rare" campaign gets big results and bigger budget https://news.centurionjewelry.com/articles/detail/one-year-later-dpas-successful-real-is-rare-campaign-gets-big-results-and-b. Accessed July 2019
41. Trilling L (2009) Sincerity and authenticity. Harvard University Press
42. Tuan YF (1980) The significance of the artifact. Geograp Rev: 462–472

Sustainable Industrialisation for Luxury Products: Manufacturers and Retailers Must Commit to Tackling Modern Slavery in Africa

Ismail W. R. Taifa

Abstract Globally, there is high pressure concerning sustainability. This requires designers, manufacturers, distributors, and consumers to have obligations of looking at sustainability tenets: social (people), economic (profit) and environment (planet). Researchers have been exploring economic and environmental issues in several sectors. It is the sustainability age where people must now look at contemporary issues in the manufacturing (production) processes of luxury goods. One of the critical problems in today's supply chain of industrial luxury products is modern slavery issues (MSIs) (or neo-slavery or contemporary slavery). MSIs mostly refer to slavery that continues to occur in private individuals, groups, institutions, organisations, companies, on engaging child labour, human trafficking, forced labour, long working hours, among other forms, in manufacturing products. This study focuses on Africa regarding manufacturing and retailing of luxury products. Africa is focused because the 2018 Global Slavery Index ranked Africa number one concerning MSIs; several African countries produce precious (valuable) metals; and, many African societies cannot notice much about how critical the MSIs are. The production processes and other sustainability issues were thus explored. The findings suggest the need for Africa to strengthen consolidative interventions to fight the diverse environment that results in MSIs.

Keywords Sustainable industrialisation · Supply chain · Luxury products · Global Slavery Index · Modern slavery · Corporate social responsibility · Manufacturer · Retailer · Africa

I. W. R. Taifa (✉)
Department of Materials, Faculty of Science and Engineering, The University of Manchester, M13 9PL Manchester, UK
e-mail: taifaismail@yahoo.com; ismail.taifa@postgrad.manchester.ac.uk

Department of Mechanical and Industrial Engineering, College of Engineering and Technology, University of Dar es Salaam, Dar es Salaam, Tanzania

I. Coste-Maniere and M. Á. Gardetti (eds.), *Sustainable Luxury and Jewelry*,
Environmental Footprints and Eco-design of Products and Processes,
https://doi.org/10.1007/978-981-16-2454-4_10

1 Introduction

Africa has fifty-four countries (sovereign states) recognised by the United Nations (UN). It is the second populated continent after Asia, with around 1.308 billion people by 2019 [59]. Almost sixty per cent of her populations are people under the age of twenty-five, making Africa the youngest continent in the world [59]. Africa is a resource-wealthy continent. Within the African continent, some people engage in the production (manufacturing) of luxury products, purchasing luxury products while others own factories that add values to the luxury products. It is not easy to estimate the total number of people engaging in luxury products within Africa.

Globally, there is high pressure concerning sustainability. This requires designers, manufacturers, distributors, and consumers have obligations of looking at the major three tenets of sustainability: social equity (people), economics (profit) and environment (planet). Many researchers have been looking around economic aspects and environmental issues in several sectors. It is now the age of sustainability; people must look at contemporary issues in the manufacturing processes of luxury commodities. The luxury industry is a deep-rooted industry worth over two hundred billion United States dollars (USD) per annum worldwide; and though profitable, the luxury products' market is softly reframed to line up with crucial and evolving developments in the sector [36].

One of the critical issues in today's supply chain of industrial luxury products is modern slavery issues (MSIs) (or neo-slavery or contemporary slavery). The term 'modern slavery' has no official definition [6, 51]. However, institutions, organisations and academicians define it in reference to slavery that occurs in private individuals, groups, institutions, organisations, companies, on engaging child labour, human trafficking, forced labour, long working hours, among others, in producing or manufacturing products and/or servicing customers [5, 16, 28, 41, 54].

Globally, many people live in modern slavery during the twenty-first century than any period of humans' history [35]. From the '2018 Global Slavery Index (GSI)', the countries with the uppermost predominance of MSIs worldwide include Burundi, Eritrea, South Sudan, Cambodia, the Democratic Republic of the Congo (DRC), Afghanistan, Pakistan, North Korea, Mauritania and Iran [65]. It is not only the Asian and African countries with the MSIs; the crime is now a global problem. For instance, the 2014 GSI indicated countries like China, Thailand, Uzbekistan, India, Bangladesh, and Russia, amongst the most affected also, whereby the European Union (EU) had an estimate of half a million slaves by 2014 [25].

Due to slavery practices being illegal, it has thus increased the crime effects and become more hidden by involved people [35]. MSIs-related crimes in the twenty-first century are difficult to both law enforcement and law-makers [8]. However, there are some initiatives in seeking human rights by filing cases to courts. For instance, Microsoft, Tesla, Google, Dell, Glencore, Zhejiang Huayou Cobalt and Apple, were among the companies identified in the case filed by the International Rights Advocates for the fourteen Congolese families in 2019 [9]. The BBC's [9] case can be associated with both the manufacturers (e.g. Glencore and Zhejiang Huayou Cobalt mining companies), whereas the rest are *retailers* or *customers*. Glencore responded that the company "does not tolerate any form of [the] child, forced, or compulsory labour" [9]. Microsoft similarly said, "it was committed to responsible sourcing of minerals and that it investigates any violations by its suppliers and takes action" [9]. Likewise, Google and Apple responded by indicating how the two companies are committed to complying with ethics in sourcing the required materials [9]. Examples of those engaging in manufacturing products (*manufacturers*) include the South Africa scenario. Here, the manufacturer was accused of involving MSIs in manufacturing products [45]. The manufacturer was found with children, among whom some minors were as young as fifteen [45].

Africa continent, with other continents, requires collaborative measures to eradicate MSIs in attaining sustainable development goals (SDGs) in each section. For example, the ILO [29, 30] report provided global statistics about forced labour exploitation by economic activities. These include 15.1% of the exploitation, domestic work (24.3%), mining and quarrying (4%), wholesale and trade (9.2%), construction (18.2%), agriculture, forestry and fishing (11.3%), personal services (6.8%), begging (0.8%), arts (0.1%), illicit (0.7%), and accommodation and food service activities (9.5%) [29]. Many of these activities involve either processing or servicing varieties of luxury products.

Although modern slavery is a global problem, this study focuses mainly on the African continent, specifically on how the manufacturing (production) processes of luxury products are performed. Why is this analytical study focused on Africa? The considered reasons include the following:

(a) The 2018 GSI ranked Africa number one in MSIs [65]. One hundred sixty-seven countries were covered in the 2018 GSI [65]. Similarly, although MSIs occur in all world's regions, by 2016 Africa had a prevalence of 7.6 per one thousand people, Asia, and the Pacific (6.1 per one thousand), and Europe and Central Asia (3.9 per one thousand). MSIs statistics must be elucidated carefully because there is not enough information in some regions, particularly the Arab States and the Americas [30].
(b) The majority of African countries produce precious (valuable) metals, including Gold, Tanzanite (only from Tanzania), Platinum, Palladium, Rhodium, Iridium, Diamond, Copper (sometimes considered as a precious metal because it makes jewellery and currency), etc. Cobalt mining activities can also be looked at because such metal makes batteries for electronic powered luxury products or brands [3]. Africa being a resource-wealthy continent, including minerals, makes the continent to have several mining activities which perhaps involve MSIs. This study focuses on the manufacturing (production) processes; however, such processes of precious minerals are mostly performed out of Africa. Most of these minerals are sold abroad in raw form. Nevertheless, South Africa is an excellent example amongst the African countries which produce Gold [27].
(c) Many African societies cannot notice much about how critical the modern slavery problem is. Maybe this is because MSIs are strangely affecting the most

downgraded members of societies [20]. Such people consider themselves less powerful to raise their concerns to their leaders and other responsible people, organisations, or relevant authorities.

Hence, the above three reasons make the need for this study which mainly explored sustainable industrialisation for luxury products with key concern on how manufacturers and retailers should assist in tackling MSIs in Africa. Four specific objectives were considered as follows:

(a) To explore MSIs in manufacturing (producing) luxury products.
(b) To explore available and implementable sustainable practices to eradicate MSIs.
(c) To explore how manufacturers and retailers can assist in tackling MSIs in Africa
(d) To propose a suitable integrative framework to increase awareness and eradicate MSIs.

2 Theoretical Background

2.1 Luxury Products

Recently there has been a rapid demand increase of luxury products [1, 14, 21, 36] than ever before probably due to the easier accessibility fostered by technological improvements in manufacturing goods, despatching and selling them. Luxury depends on regional, economic or cultural circumstances [24]. Since ancient times, luxury has always been a sign of social status, power and prosperity [24]. Serdari [46] defines luxury as "a universal certainty that implies scarcity, beauty, and culture. It is a multi-sensory experience that impacts the individual on a sensorial [or] physical, emotional, and intellectual level."

Similarly, a luxury product is an "inessential, desirable item which is expensive or difficult to obtain" [43]. Luxury products can similarly be described as "expensive things, such as jewellery and make-up, that are pleasant to have but are not necessary" [13]. Luxury products are linked with wealth, exclusivity, power, and the satisfaction of non-essential demands [13, 24, 43, 46].

Although the majority view price (affordability) as the primary criterion for luxury products [18], this is contrary to Prendergast and Wong's [44] perspective who observed that luxury brands for some people, including parents, are linked with good design and quality rather than the price (affordability). There are also factors associated with luxury products. According to Vigneron and Johnson [63], the structure of the 'brand luxury index' is either personal or non-personal perceptions. The personal ones include hedonic and extended, while the non-personal ones comprise of quality, uniqueness, and conspicuousness. The recent theoretical underpinnings suggest numerous factors that can potentially impact buyers' perception of luxury products, including factors like availability of counterfeits, country of origin, economic and the sense of belongingness [1]. Danese et al. [21] also indicate that each industry may have its critical success influences; however, for the luxury section of the fashion business, such factors include style and design, product quality, creation of a lifestyle, country of origin of goods, emotional appeal (mostly regarding the shopping experience) and brand reputation. Gardetti [23] recently also discussed the fundamental attributes of luxury—code of luxury—along with their interactions with sustainability. Such attributes included innovation; high price; uniqueness, exclusivity, and scarcity; excellent quality and *savoir fair*; and personal history, ancestral heritage, and legacy. What is the age group that purchases luxury goods? All age groups can demand luxury products; for Africa, the young generation is, however, considered to demand more luxury products than other generations.

Africa is projected to be the future luxury market. Some of the key factors driving the growth of the luxury market include the following. There has been a robust economic growth from the beginning of the twenty-first century, bolstered by high product prices, enhanced economic and political governance, and heightened foreign investment [37, 18]. Likewise, for the last ten years, Sub-Saharan Africa has become the second-quickest-developing region globally, the first being Asia-Pacific [18]. The positive development indicators include the thriving demographic, a fast escalating young population, an increase and growing wealth of the middle class, urbanisation, booming of women financial independence, growth of the hyper-rich people, and the increase of advanced technologies to foster communication and digitalisation [18]. Africa is thus projected to be the second quickest-expanding region of luxury products globally. Similarly, Sub-Saharan Africa is projected to grow its luxury markets by 30% in the next five years [18].

2.2 *Mapping Africa's Natural Resources*

Figure 1 depicts a summary of some of Africa's primary natural resources. These resources indicate how Africa is rich with numerous treasured resources. Nevertheless, the wealth creation in Africa is not uniformly distributed, and in many countries with the largest economies, larger overall wealth is owned by a small population [18]. The availability of many resources suggests the need for sustaining them with clear and implementable robust policies. Many resources (Fig. 1) result in the production of luxury products. As the world is now focusing on sustainable luxury products together with sustainable industrialisation; thus, manufacturers and retailers, with other stakeholders, are now receiving high pressure to achieve *sustainability*. The most sustainability aspect directly affecting society is *social equity values*. This has emerged with corporate social responsibility (CSR). Within CSR as the international private business self-regulation, researchers are now investigating MSIs in each step of the manufacturing and retailing of the basic and luxury products. This research thus focuses on luxury products concerning the realisation of sustainable industrialisation by manufacturers and retailers.

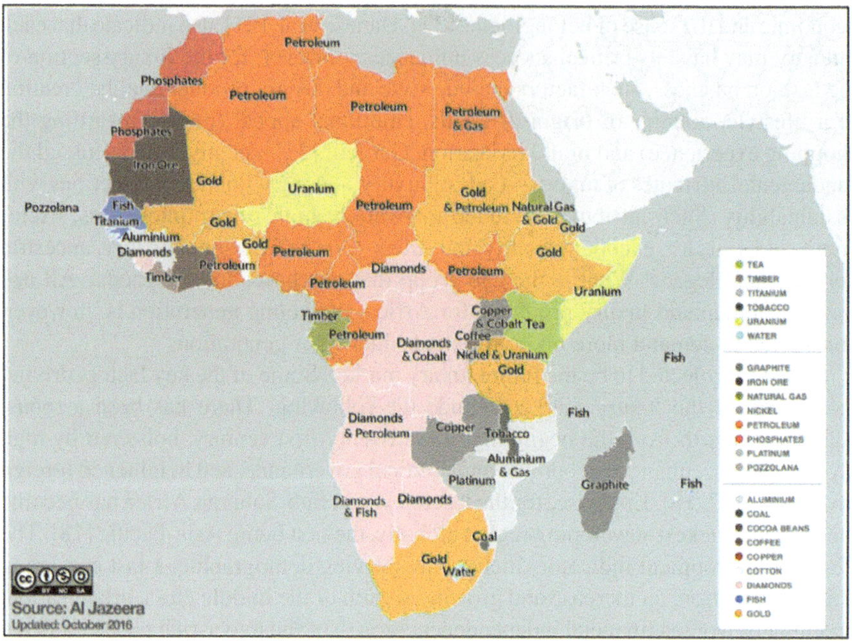

Fig. 1 A simple map for African's natural resources. *Note* The resources were labelled for each country based on what are the most valued exports. *Source* Al Jazeera [2]

2.3 *Luxury Products' Manufacturing Processes*

Both manufacturers and retailers for luxury products can be categorised as micro, small, medium, or large enterprises. Both retailers and manufacturers can involve directly or indirectly into the supply chain (SC) of luxury products. Considering that the manufacturing process is part of the general SC of luxury goods; several luxury products may thus have different SCs, but the general framework remains the same. Production is the focal point of businesses; however, it has two contradictory outcomes: distributing essentials and facilities for humanity whilst diminishing the worldwide commons which support life [66].

Figure 2 illustrates the general example of the SC of luxury products which include the manufacturing process. The SC in Fig. 2 includes raw materials, the manufacturing premises (e.g. firm, factory, or plant), the distribution processes, retailers, and the retailing processes. Similarly, the chain comprises information sharing, cash flow, order flow and reverse logistics directions. All nodes of the chain require to be sustainable: the sustainability tenets must be well-achieved.

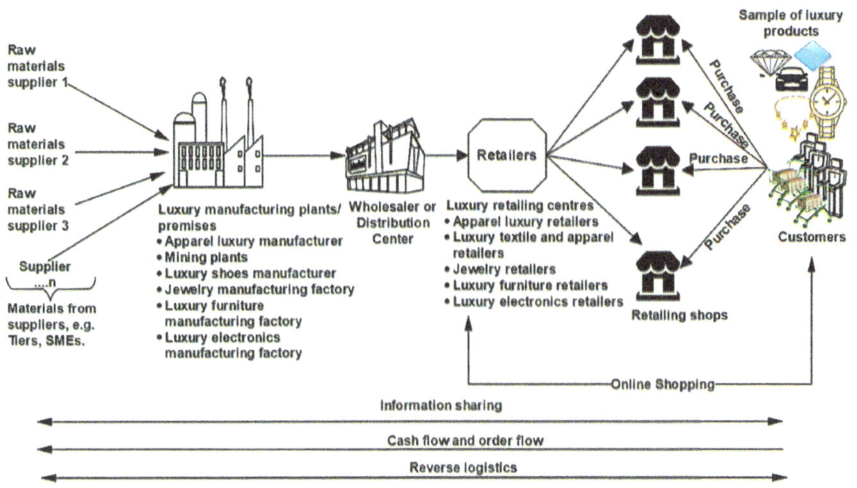

Fig. 2 A generic example of the supply chain for luxury products. *Source* Created by the author

2.4 The Global Scenario on the Hidden and Everyday Problem: Modern Slavery

The twenty-first century has seen the emergence of high pressure from both the government and societies towards eradicating modern slavery issues (MSIs). NHS England defines MSIs as "the recruitment, movement, harbouring or receiving of children, women or men through the use of force, coercion, abuse of vulnerability, deception or other means for the purpose of exploitation."[1] MSIs are now global concern as it affects both intranational and international [53]. MSIs affect many people, through their societies or at their working places. Several sectors are affected, including the manufacturing [21, 34], public health [15, 38], mining [3, 42, 62], etc.

Consistent with Walk Free Foundation [65], among the G20[2] members, seven countries—the UK, the USA, Italy, Germany, France, China and Brazil—are taking actions while the other twelve countries are not taking actions—South Africa, Japan, Turkey, India, Mexico, Indonesia, Russia, Saudi Arabia, Argentina, South Korea, Australia,[3] and Canada. The top five categories of products which are at risk of contemporary slavery practices—MSIs—imported into the G20 (by USD) include mobile phones, computers and laptops (USD 200.1 billion); sugarcane (USD 2.1 billion); clothes (apparels) (USD 127.7 billion); cocoa (USD 3.6 billion); and fish

[1] Modern slavery definition by NHS England is at https://www.islingtonccg.nhs.uk/YourHealth/modern-slavery-and-human-trafficking-statement.htm (accessed on 09 May 2020).

[2] See the G20 members at https://g20.org/en/about/Pages/Participants.aspx (accessed on 27 April 2020).

[3] Australia was expected to introduce "transparency of supply chain" by the middle of 2018 [65]. Visit at https://www.globalslaveryindex.org/2018/findings/highlights/.

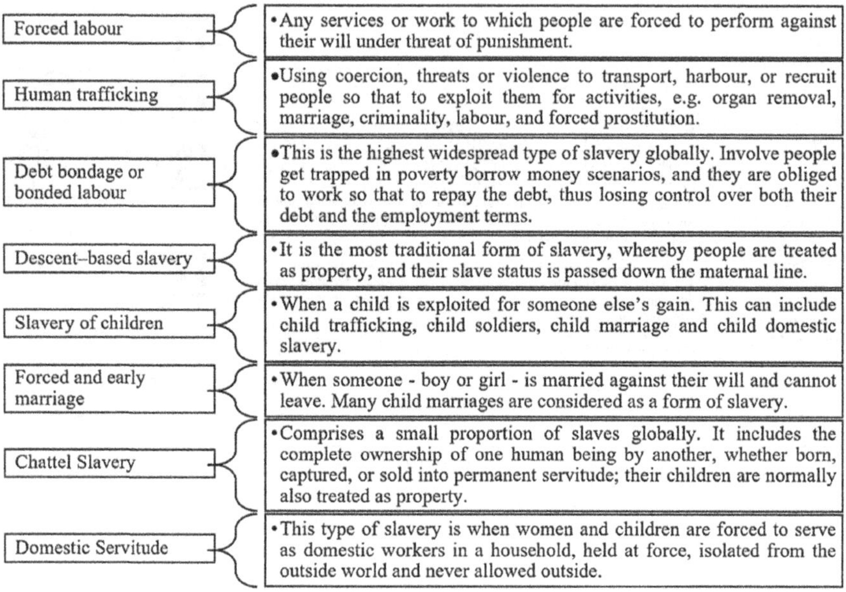

Fig. 3 Types of modern slavery in society. *Source* Boddy-Evans [10] and Anti-Slavery International [5]

(USD 12.9 billion) per annum [65]. The estimates for MSIs were conducted based on one hundred and sixty-seven countries.

Considering that the UN recognises 195 sovereign countries[4]; it is thus essential to indicate how many products are at risk of MSIs and what the cost is for each country. Africa has fifty-four countries; the key question thus should be on how many basic and luxury products are manufactured, exported, and/or consumed within the continent? The estimated MSIs statistics indicate the need to take action to stop sourcing products and services from manufacturers who are showing signs of slavery practices. The G20 countries should lead such actions as they spent more than other countries. Up to this date, researchers have reported many modern slavery types globally (Fig. 3). According to ILO [29], the MSIs are mostly categorised as forced labour and forced marriage while the Anti-Slavery International [5] and Boddy-Evans [10] further classify MSIs, as shown in Fig. 3.

Due to the emergence of diverse and new forms of slavery, there are increased further challenges to generate reliable estimates of modern slavery crimes [35]. Both the true size of the total luxury manufacturing industries (firms) and primary slavery statistics are not precisely known since only the data through international institutions are mostly referred by many researchers, especially in Africa. Available statistics on the severity of MSIs differ widely [8]. By 2016, there were over 40.3 million

[4] The UN permanent members are 193 while the Holy See and the State of Palestine are non-member observer states. See the details at https://www.un.org/en/member-states/ (accessed on 08 May 2020).

victims—children, women and men—of modern slavery, whereby 71% were female, and 29% were male [65]. Of these, 24,900,000 (62%) were forced labourers while 15,400,000 (38%) were forced marriages [65].

The 'forced labour' means they were forced to work contrary to their will while the 'forced marriage' means such people, mostly women and young females, were forced to be married to people they had not agreed upon. Further categorisation is that, of the 24.9 million (forced labourers), over 16 million are from domestic work, agriculture, and construction, leaving the other 8.9 million for other sectors including the manufacturing sector [54]. Nowadays, all sectors should disclose information linked to the provenance, environmental, supply chain memberships (partnerships), social, among other information [39, 50].

The International Labour Organisation (ILO) is leading to eradicate MSIs [29]. To achieve SDGs successfully, this must include exterminating all MSIs. Such achievements create added necessity to understanding the magnitude and manifestations of these severe crimes. Essential measurements must be evaluated through accurate decisions concerning the appropriate policies, resources, and necessary consolidative interventions. All fighting measures against the MSIs face challenges as this process involves the investigation of the deeply hidden criminal activities through the daily operations in all sectors. Thus far, researchers have been adapting the global estimates statistics from the ILO. Such statistics are not clearly stated for each sector nationally. Some of the available ILO's [28–30] data regarding MSIs are in Figs. 4 and 5.

2.5 *Sustainable Industrialisation for Luxury Products*

Luxury products are regularly followed by pleasant, exclusive, attractive, expensive, and beautiful features [36], contrary to sustainability which is mainly defined based on the three tenets: social (people), environment (planet) and economic (profit) (Fig. 6). The two terms—luxury and sustainability—are not directly related. Despite the differences between the two terms, the world is facing high pressure concerning sustainability. Such pressure builds up from within the specific industries, consumers (customers), from non-related industries, and international organisations, including the United Nations (with all associated departments). Recently, researchers have been combining luxury and sustainability to refer to 'sustainable luxury' [24, 36]; thus, leading to sustainable manufacturing or sustainable industrialisation (in the bigger picture). This makes designers, manufacturers, distributors, and consumers have obligations of looking at all sustainability tenets.

To enhance the sustainable manufacturing processes for luxury products, such processes should integrate the efficiency and profit-oriented goals with comprehensive enterprise's both internal and external stakeholders, together with consideration of its social and environmental impacts to the current and future generation. Internal and external stakeholders' consideration means enterprises must pay attention to the entire supply chains of the specific luxury products. Also, both luxury

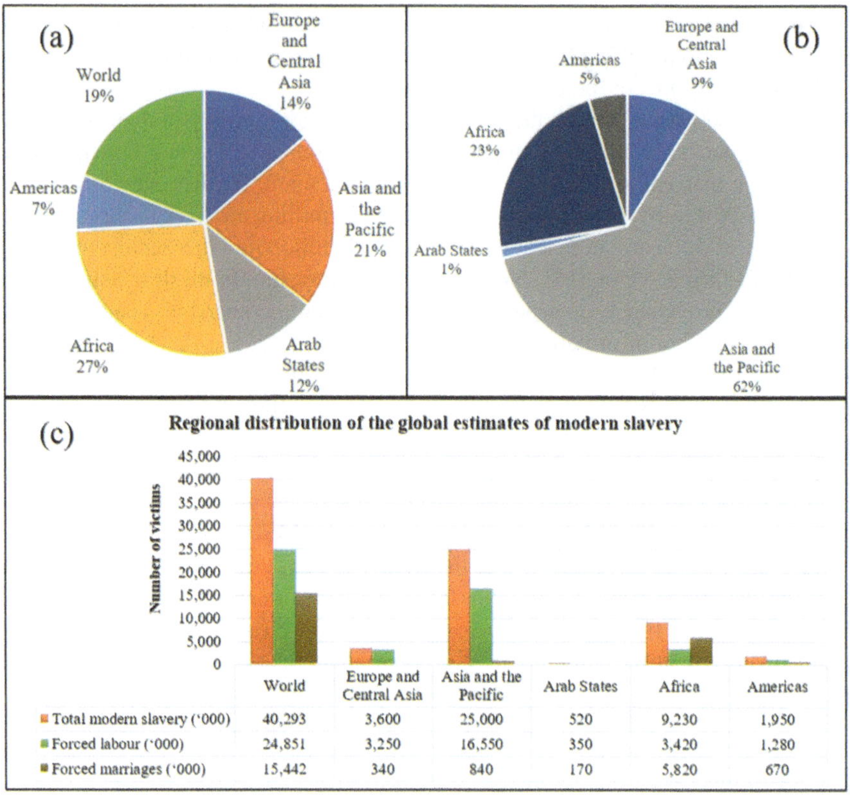

Fig. 4 Estimated MSIs data. **a** Prevalence of MSIs by region (total victims per one thousand people). **b** Regional distribution of MSIs ('000). **c** Regional distribution of the global estimates of MSIs. *Source* ILO [29]

products manufacturers and retailers must show implementable initiatives and interventions. The initiatives and interventions should also be supported with dedicated resources and robust policies of the international organisations (or institutions) and the country's specific policies to eradicate MSIs which are involved in manufacturing luxury products. There is no easy approach that can enhance sustainable industrialisation for luxury products without the full involvement of manufacturers and retailers in tackling MSIs in Africa.

Although sustainability is attributed to three tenets (Fig. 6), this study mostly focused on social responsibility (also identified as CSR). World Business Council for Sustainable Development described CSR as "the continuing commitment by business to behave ethically and contribute to economic development while improving the quality of life of the workforce and their families as well as of the local community and society at large" [40].

ILO [31] also describes CSR as "a way in which enterprises give consideration to the impact of their operations on society and affirm their principles and values

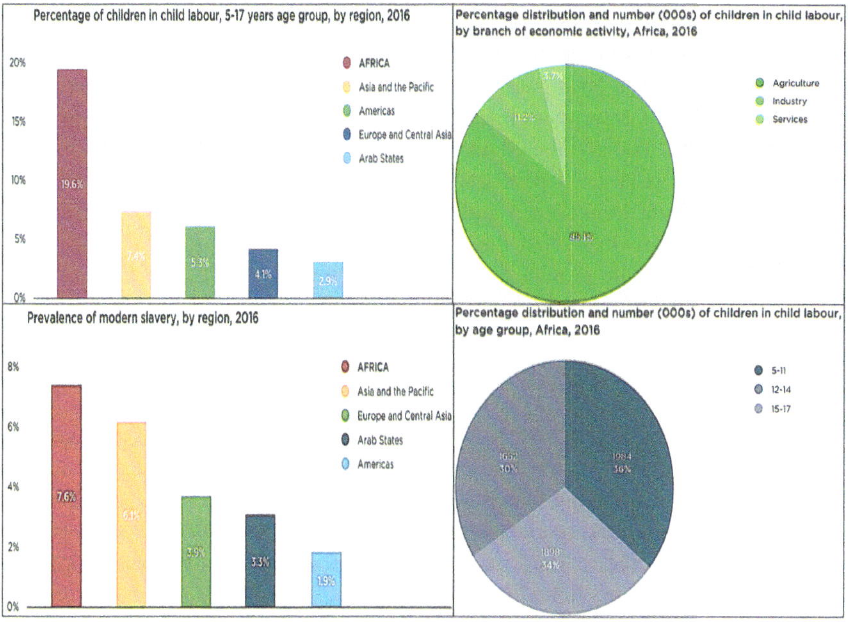

Fig. 5 Estimated modern slavery statistics in Africa. *Source* ILO [28]

Fig. 6 Sustainability tenets. *Source* Developed by the author

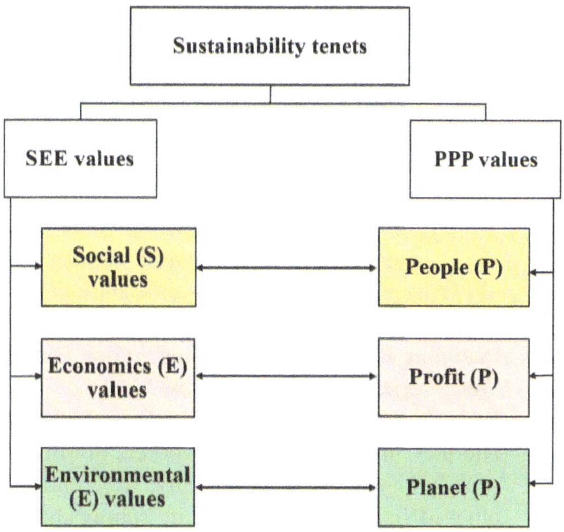

both in their own internal methods and processes and in their interaction with other actors." CSR has been predominant for many decades, and consequently, it has had an enormous influence on organisations' performance [64]. Currently, there is an escalating number of enterprises information about CSR to measure the influence of their actions on the society, the environment, economy, and communicate this with their stakeholders [64]. CSR can be viewed by five aspects: commitment, conviction, communication, consistency and credibility [47]. Africa should desire to have more sustainable growth so that the super-rich consumers do not drive the luxury market [18]. Sustainable growth should foster economic, environmental, and social responsibility values.

According to CPP Luxury [18], in 2017, the African luxury sector created around six billion in revenue (USD). This revenue comprises private jets, luxury watches, luxury apparel and accessories, luxury cars and vehicles, yachts and luxury lodges and hotels. By 2017, South Africa, Kenya, Nigeria, Morocco, and Egypt, were the leading luxury markets by revenue [18]. Such revenue was enhanced by a large luxury hotel sector in the mentioned countries. Other luxury products can be luxury shoes, bags, furniture, luxury electronic gadgets, drinks, food, flowers, construction equipment, among others. There is a possibility for some of these products to be manufactured by workers linked in MSIs.

Most of these luxury products and services involve supply chains. For consumers, it is not an easy and successful process for them to question the manufacturing processes. Since it is the sustainability era where both stakeholders and shareholders are accountable to eradicate MSIs, do customers (consumers) ask whether there are MSIs involved within the entire supply chain? Amnesty International [3] asked an interesting question that "*Is my phone powered by child labour*?" This is because there is evidence that the cobalt, which is the major source for phones' batteries, is associated with child labourers in the DRC [4, 9]. Some of the questions that can further be asked by customers (consumers) include:

(a) Who made luxury products bought and used by customers?
(b) Were slavery practices involved in manufacturing such luxury products? Is it possible for customers to know the initial (exploration processes) to the final personnel (selling processes) who were involved in processing Gold, Tanzanite, Platinum, Palladium, Rhodium, Iridium, Diamond, Copper, etc.?
(c) Were workers paid appropriately?
(d) Were they exploited as bonded labour, forced labour, and the like?
(e) Did they involve vulnerable workers in offering services and/or manufacture such luxury products? Generally, were children, older people, refugees, and other MSIs involved in the entire supply chain to process luxury products?

Since some supply chains, including the one for the textiles and apparel sector [55], mining sector, among others, are dynamic, long, and complex, it is challenging to notice any form of slavery quickly. There are limited theoretical and empirical studies on MSIs in several sectors, including the business and management aspects [17, 19]. For the general supply chain viewpoint, researchers explored MSIs for several products [25, 34, 41, 42, 62]. Other researchers extended their investigation by looking at the global value chains regulated by multinational companies [51]. Hardly any study looked at MSIs in luxury manufacturing focusing on Africa, specifically how manufacturers and retailers can foster the eradication measures.

Also, manufacturers of any product nowadays are emphasised to collaborate as an extended enterprise [55, 56]. Such collaboration necessitates selecting manufacturers or companies to work with. As identified by Taifa et al. [54], when selecting manufacturers for both the luxury and basic products, nowadays it is essential to consider MSIs and other CSR-related issues. This also applies to retailers who look to sources from their suppliers (vendors) or manufacturers.

3 Methodology

Qualitative data were collected, mostly from secondary sources. The production process was analysed together with the value chain management and other sustainability issues on supply chain management. All fifty-four countries were analysed to find out the documented information about MSIs. The analysis comprised the following checklist:

a) Which African countries are linked with MSIs?
b) Which sectors are most affected by MSIs?
c) Amongst the affected sectors, is mining included? What about other luxury products?
d) Any presence of the modern slavery statement?
e) What are the initiatives in combating MSIs?
f) Are there practices to evaluate and compare companies' CSR performance?
g) Have African countries enacted laws to eradicate MSIs?

To accomplish an analytical approach, a methodological decision analysis model (MDAM) was employed (Fig. 7). MDAM helped to set exclusion and inclusion criteria.

4 Generalised Results

4.1 Modern Slavery Information Within Africa for Luxury Products

The results indicate that there are no statistics to assess and contrast companies' CSR execution for the specific supply chains of luxury products. This is supported by Virtanen [64] that the CSR reports are regularly inappropriate and lack quantitative data for many companies that users—stakeholders and shareholders—could use to compare the CSR performance. Through the use of the MDAM (Fig. 7), it was also

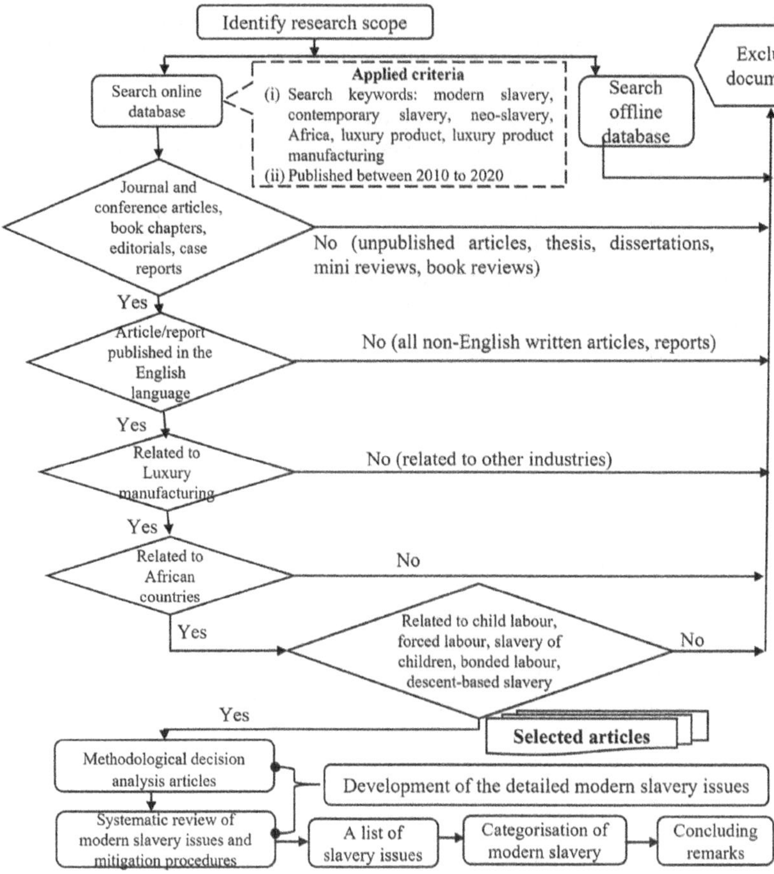

Fig. 7 Methodological decision analysis model *Source* Developed by the author

not easy to find researchers who investigated MSIs in connection to luxury products industrialisation in Africa. The available data are based on general MSIs, and others are accusations which are not supported by evidence. Based on Fig. 7, some forms of slavery practices were identified concerning luxury products exploration and/or manufacturing tasks within some African countries. Table 1 depicts the summarised data for luxury products. For example, the findings from Amnesty International [3] indicate that there are over "40,000 children [who] work in southern DRC where the cobalt is mined, [thus,] there is a chance that our phones contain child labour." This is a terrifying finding, as some children are as young as seven to fifteen years old. Obviously, some information related to slavery practices in Table 1 were still allegations which were yet to be proven as the cases are still at courts. Nevertheless, such allegations indicate the symptoms of having slavery practices within the industrialisation of some luxury products within Africa. Furthermore, the 2018 GSI found that the North African nations had the highest rates of MSIs with nearly four

Table 1 The slavery forms concerning luxury products exploration and/or manufacturing within the few selected African countries

African countries	Slavery forms in manufacturing luxury products	General perspectives	Hint to eradicate MSIs in manufacturing, producing, or servicing	Source of information
Chad	Child labourers, human trafficking, forced labourers in gold mines	There are reports regarding Chad being the source, transport, and destination for victims: child trafficking. Children engage in forced labour in several industries and farming activities. Also, several migrants worked in gold mines to pay for their journeys	Both the government and other institutions should lead the means to eradicate MSIs. Manufacturers and retailers who benefit from the mined gold in Chad must commit to abolishing all slavery practices	Freedom House [22]; Jesperson [32]
South Africa	Human trafficking, child labourers (around the age of 15), forced labourers, manufacturers paying low wages (below the national minimum wage), in the T&A industry	The manufacturer was accused of involving some forms of slavery activities in manufacturing products. The manufacturer was accused of unlawful deductions from workers' wages	Here, the government accused the manufacturer. This is a good step as the court was involved too. Still, this should lay a fundamental necessity of having strict laws to all manufacturers regarding slavery practices	Steele [49]; SAnews.gov.za [45]

(continued)

Table 1 (continued)

African countries	Slavery forms in manufacturing luxury products	General perspectives	Hint to eradicate MSIs in manufacturing, producing, or servicing	Source of information
Sudan	Trafficking victims, including women and men, are forced to operate in gold mines near Dongola around the Al Khandag	The victims for gold mining activities came from Sudan, DRC, South Sudan, Chad, Eritrea, and Niger. Some people are forced to work in gold mining areas to raise funding to pay the ransom for the trafficked people	Both stakeholders who benefit from gold and other luxury goods, should lead in combating slavery issues in Sudan. Sudanese authorities have been taking appropriate actions to punish human traffickers severely	Krueger [33]; Ati [7]
DRC	There are damages over injuries and deaths of child labourers in mining cobalt minerals whereby some children were as young as 7 to 15 years old	Globally, 60% of the supply of cobalt is from the DRC. There are over 40,000 children in cobalt mines. Cobalt makes lithium-ion batteries for charging smartphones, electric automobiles, computers, and laptops	The mining companies, manufacturers, retailers, customers (consumers) can interactively help to eradicate these practices. The key question should be "*Is my phone, laptop, computer, car, powered by child labour?*"	Sovacool [48]; BBC [9]; Amnesty International [3]

(continued)

Table 1 (continued)

African countries	Slavery forms in manufacturing luxury products	General perspectives	Hint to eradicate MSIs in manufacturing, producing, or servicing	Source of information
Ivory Coast	Unpaid family labourers, human trafficking, child labourers and slave labourers	Cocoa is a crucial ingredient for chocolate making. High-quality chocolate is considered as a luxury product because society describes it as an excellent exclusive for the highest social status of people	The cocoa industry in West Africa and Ivory Coast should tackle any slavery practice. All countries that export and import chocolate or cocoa should also collaborate with the Ivory Coast government and consumers to eradicate such crimes	Crane [19]; Christ et al. [16]
Madagascar	Child labourers were used in the mines— as young as eight—because they could enter the cramped areas more effortlessly than matured workers	The gem industry involved child labourers due to the discovery of sapphires in Madagascar	The Madagascar government enacted a law to prohibit involvement of child labourers below 15 years old and set the minimum age for employment as well	United States Department of State [61]; O'Driscoll [42]; van der Wal [62]
Ghana	There is an involvement of child labourers in cocoa farming activities. Chocolate, as the end product of cocoa, is still exclusive to the higher social classes, thus making it a luxury product	Between 2013 and 2017, children were forced to work in cocoa farming by other people apart from their parents. There was an increase of twenty child labourers per 1,000 children employed in cocoa agriculture	Chocolate manufacturers and retailers are obliged to contribute their consolidative efforts in eradicating slavery practices in all cocoa agricultural-related activities in Ghana. Joint efforts and programmes are required	O'Driscoll [42]; Walk Free Foundation [65]

hundred sixty-five thousand people captured and subjected to exploitation globally, i.e. one in eighty of her population [65].

Furthermore, through secondary data, there is a global list of products associated with forced and child labourers as of 2016. Table 2 summarises some African countries to whom the manufacturing or processing chains of their luxury products are associated with MSIs. This further proves the need for having consolidative eradication methods of MSIs within Africa. In particular, there would not be achievements regarding sustainable industrialisation of luxury products within Africa without setting implementable policies. Nowadays, there has been an increase in slavery activities in several operations.

4.2 Modern Slavery Act (MSA) Statement in African Countries

According to the World Gold Council [67], the "Modern Slavery Act ('the Act') is a significant step in confronting human trafficking and modern slavery, both in the UK and abroad." There is a need for African Union (AU) to initiate 'Africa Modern Act Slavery Policy' as a means of pushing forward for reformation within the continent. This should be in line with the necessary instructions for all the AU fifty-five[5] members to enact laws regarding MSIs within their countries. The laws to be enacted should prevent both manufacturers, retailers, among other stakeholders and shareholders, in combating MSIs to ensure sustainable industrialisation. All companies must be obliged to include the MSA statement on their websites. An inclusive of the MSA to companies' websites shows supply chain transparency. This means, no matter what the original country of registration is—whether the particular company is entirely an African company or external company—all should be fully responsible in combating MSIs in their entire manufacturing processes, retailing and other associated practices.

5 Discussion

5.1 General Discussion

The theoretical underpinnings indicate the increase of slavery practices over three times during the twenty-first century than between the thirteenth and fifteenth centuries. For Africa being the rich-wealthy continent, this can also influence the increase of MSIs. It is thus essential to eradicate MSIs in Africa. African countries

[5] It should be noted that the African Union (AU) has 55 members contrary to the 54 UN recognised sovereign countries https://au.int/en/member_states/countryprofiles2 (accessed on 24 April 2020).

Table 2 List of commodities associated with child labourers and forced labourers for African countries by 2016

African countries	All categorised products which are associated with slavery practices	Only products associated with child labour	Only products linked with forced labour
Angola	Diamond	Diamond	Diamond
Benin	Cotton, Crushed Granite	Crushed Granite, Cotton	Cotton
Burkina Faso	Cotton, Gold, Granite	Cotton, Gold, Granite	Cotton, Gold,
Cameroon	Cocoa	Cocoa	x
Central African Republic	Diamond	Diamond	x
Ivory Coast	Cocoa	Cocoa	Cocoa
DRC	Gold, Cassiterite, Copper, Diamonds, Coltan, Heterogenite, Wolframite	Copper, Heterogenite, Diamonds, Gold, Wolframite, Cassiterite, Coltan	Gold, Wolframite, Cassiterite, Coltan
Egypt	Limestone, Cotton	Limestone, Cotton	x
Ethiopia	Hand-woven textiles, Gold	Hand-Woven Textiles	x
Ghana	Gold, Cocoa	Cocoa, Gold	x
Guinea	Cashews, Cocoa, Diamonds, Gold	Diamonds, Gold, Cocoa, Cashews	x
Kenya	Miraa, Sisal, Tobacco	Miraa, Sisal, Tobacco	x
Liberia	Diamonds, Rubber	Diamonds, Rubber	x
Madagascar	Sapphires, Vanilla, Stones	x	x
Malawi	Tobacco	Tobacco	Tobacco
Mali	Cotton, Gold	Cotton, Gold	x
Mozambique	Cotton	Cotton	x
Niger	Gypsum, Gold, Trona	Gypsum, Gold, Trona	x
Nigeria	Gold, Cocoa, Granite, Gravel	Gold, Cocoa, Gravel, Granite	Cocoa, Gravel, Granite
Senegal	Gold	Gold	x
Sierra Leone	Cocoa, Diamonds, Coffee, Granite	Cocoa, Diamonds, Coffee, Granite	Diamonds
Sudan	Gold	Gold	x
Tanzania	Tobacco, Tanzanite, Sisal, Nile Perch, Gold, Coffee, Cloves	Tobacco, Tanzanite, Sisal, Nile Perch, Gold, Coffee, Cloves	x
Uganda	Coffee, Gold, Stones, Tobacco, Vanilla	Tobacco, Gold, Coffee, Stones, Vanilla	x

(continued)

Table 2 (continued)

African countries	All categorised products which are associated with slavery practices	Only products associated with child labour	Only products linked with forced labour
Zambia	Tobacco, Stones, Gems, Cotton	Tobacco, Stones, Gems, Cotton	x

Notes (**a**) Only raw materials or products which are associated with luxury goods were considered in this list. (**b**) *X* indicates that there are no associated luxury products. *Source* Summarised from the Bureau of International Labor Affairs [11]

should investigate to rescue vulnerable people who are linked with MSIs in manufacturing or servicing luxury products. This process can begin by asking the same question as Bales et al. [8] that how many victims are in modern slavery? This initiative might provide a clear picture for African countries to take tough decisions in combating this crime.

It is also possible for the majority of luxury products consumers not to be aware that many products are manufactured through the supply chain, which involves slavery activities. This is possibly due to the complex and dynamic nature of supply chain nodes, including raw materials suppliers, producers or manufacturers, distributors, wholesalers, retailers, consumers, and reverse logistics service providers, among other channel members involved. For instance, when customers purchase luxury products, such as gold wristwatches, diamond rings, necklaces, smartphones, laptop, there is no possible technology or process for customers to track any kind of MSIs in the production, delivery, and sales of such products.

Today, there is awareness concerning the eradication processes of MSIs in all sectors, including the manufacturing industries. Thus, an integrative framework with internal and external stakeholders to eradicate MSIs was proposed (see Fig. 8). Although the focus is on the manufacturers and retailers, the entire supply chain of luxury products must be evaluated and monitored. Internal evaluators are not enough to perform the work. This necessitates the need for external institutions or organisations, including the UN-related institutions and departments.

As a sustainability concern in modern slavery forms, the manufacturers and retailers must be responsible for tracking all forms of MSIs. Nevertheless, their initiatives must not be only to their indoor activities; they should track from their suppliers, corresponding to each supply chain node. This means the manufacturers must eradicate all MSIs within their production process and their suppliers of luxury raw materials. If they are distributors, they should similarly track all forms of slavery to their suppliers. Thus, the suppliers of each supplier should be strictly monitored. In today's sustainability issues, modern slavery should also involve reverse logistics service providers. For example, there are times when consumers decide to return the luxury products to either the manufacturer or other dedicated companies for either reuse, recycle or remanufacture the specific products. The MSIs must also be

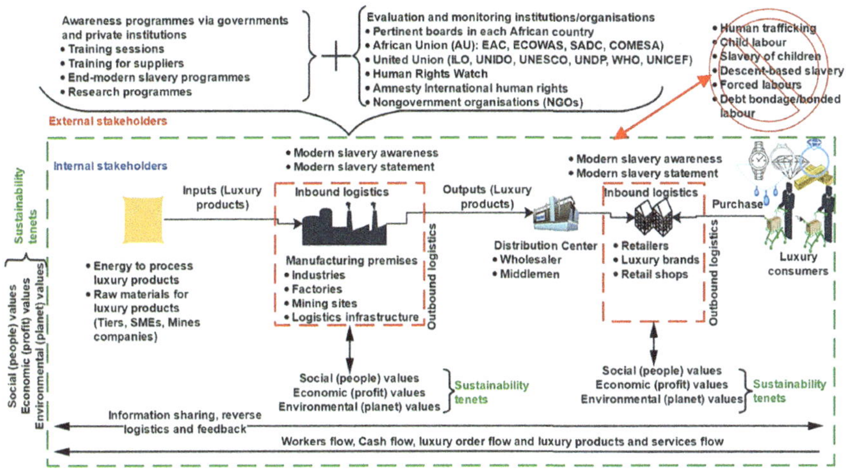

Fig. 8 An integrative framework to increase awareness and eradicate modern slavery practices. *Note* "The East African Community (EAC), The Economic Community of West African States, also known as ECOWAS, The Southern African Development Community (SADC), The Common Market for Eastern and Southern Africa (COMESA), The United Nations Industrial Development Organization (UNIDO), The United Nations Educational, Scientific and Cultural Organization, The United Nations Development Programme (UNDP), The World Health Organization (WHO), The United Nations Childen's Fund (UNICEF)". *Source* Created by the author

tracked and strictly be eradicated to such form of logistics, rather than leaving only the initiatives within the outbound and inbound logistics.

Also, the supply chain through the need for a short lead time of bulk products from a small luxury company can lead to some sort of MSIs. For example, if a retailer places an order to a company—a luxury products manufacturer—with the requirement to deliver the product in the short lead time. The manufacturer can decide to outsource the order to other companies to meet the retailer's requirements of short lead time. By doing so, the primary manufacturer cannot clearly track all forms of modern slavery in manufacturing, polishing, or producing and delivering the products to the retailers. Through such a simple example, both stakeholders—luxury manufacturers and retailers—are accountable in a case that the MSIs are tracked in either producing, delivering (transporting) or selling the particular luxury products. Hence, to eradicate such processes, both entities must show commitment to eradicating any slavery form. Their initiatives must be evaluated and monitored by both the UN, AU, countries' related institutions (departments) in consideration with all key policies and strategies to eradicate MSIs.

Furthermore, Africa requires to make task forces that would help to fight against the MSIs. The AU can set robust policies to eradicate MSIs. They should emphasise for each AU member to form task forces. MSIs require integrative measures. However, each government must lead the fight. The good example was stated by the former British Prime Minister, the Rt Hon Theresa May who said: "Modern slavery is an abhorrent crime that denies its victims of liberty, and it is disturbing to think that some of the products we buy could have been produced by someone exploited into forced labour. As global leaders in the fight against modern slavery, (…) this will not be tolerated in the UK—and our consumers will not stand for it either" [26]. MSIs destroy people's lives as they victimise, brutalise and exploit the most vulnerable globally [58]. The victims of domestic exploitation, sexual exploitation, human trafficking and forced labour, cower plainly in many towns, communities and businesses [58]. So, such practices should not have places and acceptance in any society globally.

Joint programmes are also needed to get rid of MSIs in the luxury products supply chain. One example includes the 2017 South Africa programme, which was funded by the UK. The programme aimed to shape the scheme that offers specific manufacturers and countries with specific guidance on deterring, detecting and dealing with forced labour practices [52]. The programme also helped "peer-to-peer" communication and dialogue and learning throughout the worldwide supply chain [52]. Such business collaboration initiated to eradicate forced labourers in wine and fruit-producing countries, including South Africa. Participants and other stakeholders were recommended to be provided with the best practice toolkit, information posters, leaflets for employees, guidance, checklists, template documents, and short awareness-raising film [52]. Over ninety-four per cent of more than one hundred participants to date acknowledge how the joint programme increased their understanding concerning the robust measures to be undertaken to eradicate forced labour in their businesses [52]. The other several programmes to eradicate MSIs are listed by the UK HM Government [58], with the allocated funding. For instance, Africa was allocated £10 million (2019-2026) to run the "Action of Children's Harmful Work in African Agriculture" programme [58]. Another programme is "Stamping Out Slavery in Nigeria", which was allocated £10 million between 2018 and 2023 [58]. Other profited African countries from the UK's programmes include DRC, Ethiopia, Mauritius, South Africa, Sudan, and Ghana.

Figure 9 illustrates a collaborative concept on how African governments, luxury businesses and consumers can engage in eradicating MSIs in sustaining luxury manufacturing and the retailing processes within Africa. It must be noted that MSIs require multiple solutions because there is "*no one size fits all*" procedure for tackling the conundrum [6]. Figure 10 depicts primary steps that can be used to monitor and control of modern slavery eradication policies by both stakeholders: luxury products manufacturers and retailers. Considering that the MSIs are very disguised, and the companies involved have huge financial muscles; such steps might thus not abolish the problem in question. Nevertheless, using such steps, together with joint programmes which are enforced by strict laws, the MSIs can be eliminated gradually, and in the long run, the problem might be entirely abolished.

Moreover, the global population increases exponentially every year. The population fosters the demand increases in luxury products. This, as well, should accelerate sustainable production projects which stimulate the establishment of manufacturing

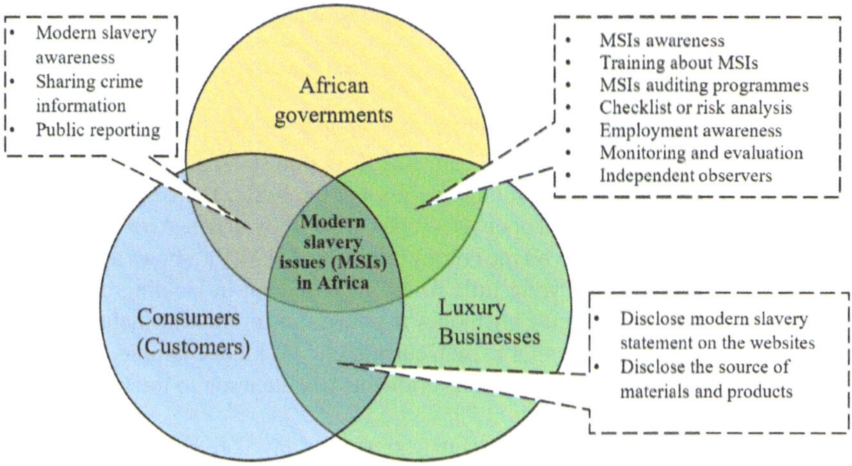

Fig. 9 An integrative concept for eradicating modern slavery in Africa. *Source* Developed by the author

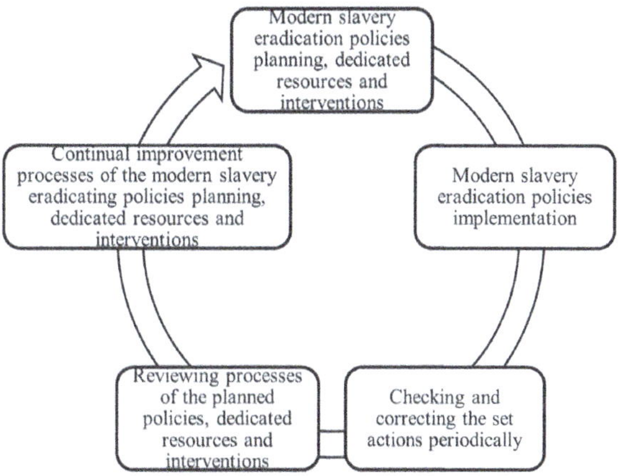

Fig. 10 A simple monitoring and evaluation process of modern slavery eradication planning. *Source* Developed by the author

systems which lead increased efficiency and productivity, while concurrently benefiting the environment and society [66]. Sustainable industrialisation for luxury products through the abolition of modern slavery foster achievement for of the SDGs, including SDG 9—"Build resilient infrastructure, promote inclusive and sustainable industrialisation and foster innovation"; and SDG 11—"Make cities and human settlements inclusive, safe, resilient and sustainable" [60].

5.2 Corruption Perceptions Index (CPI) Versus Global Slavery Index (GSI)

Due to the sensitive nature of the MSIs, there is a high possibility of corruption practices that also either prevent eradicating MSIs or punishing giant companies and organisations which might be currently involved in the MSIs. The CPI can thus also be linked to the GSI. This could open the scope of this study to non-African countries so that this could be diluting a bit the colony tune. The 2019 CPI "shows a staggering number of countries are showing little to no improvement in tackling corruption. [CPI] also suggests that reducing big money in politics and promoting inclusive political decision-making are essential to curb corruption" [57]. For example, corruption is more prevalent in countries where governments pay attention to the influences of rich people or organisations [57].

The 2019 CPI used a scale from 0 to 100; 0 is highly corrupt, and 100 is very clean. Figure 11 presents the global 2019 CPI for 180 countries around the world. The average 2019 CPI for African Union is 32/100, Americas (43/100), Western Europe and European Union (66/100), Sub-Saharan Africa (32/100), Middle East and North Africa (39/100), European Union (64/100), Eastern Europe and Central Asia (35/100), Asia Pacific (45/100) and Arab states (34/100) [57].

Whilst manufacturers and retailers should engage fully in tackling MSIs, corruption was also determined amongst the key factors that require close attention. For example, "cobalt benefits and motivates some of the largest corruption networks in Congo" [12]. Extensive ideas on MSIs are revealed with corruption, bribery and human rights issues dominant [16]. It is thus crucial dealing with corruption and opaque business connections to combat MSIs [12].

6 Concluding Remarks and Recommendations

6.1 Conclusion

Both manufacturers and retailers for luxury products can either be micro, small, medium, or large enterprises. Any form of enterprise is at risk of either producing luxury products or to be supplied with products linked to slavery practices. This chapter discussed sustainable industrialisation for luxury products through efforts from luxury manufacturers and retailers on how they should commit to tackling MSIs in Africa. Although MSIs are global, complex, and hidden, this study focused on Africa with crucial aspects of how luxury products manufacturing processes should not involve MSIs. Some luxury products involve long, dynamic, and complex supply chains, thus leading to an increase of risks of MSIs by manufacturers and retailers. Other luxury products like precious minerals exploration, such as Gold,

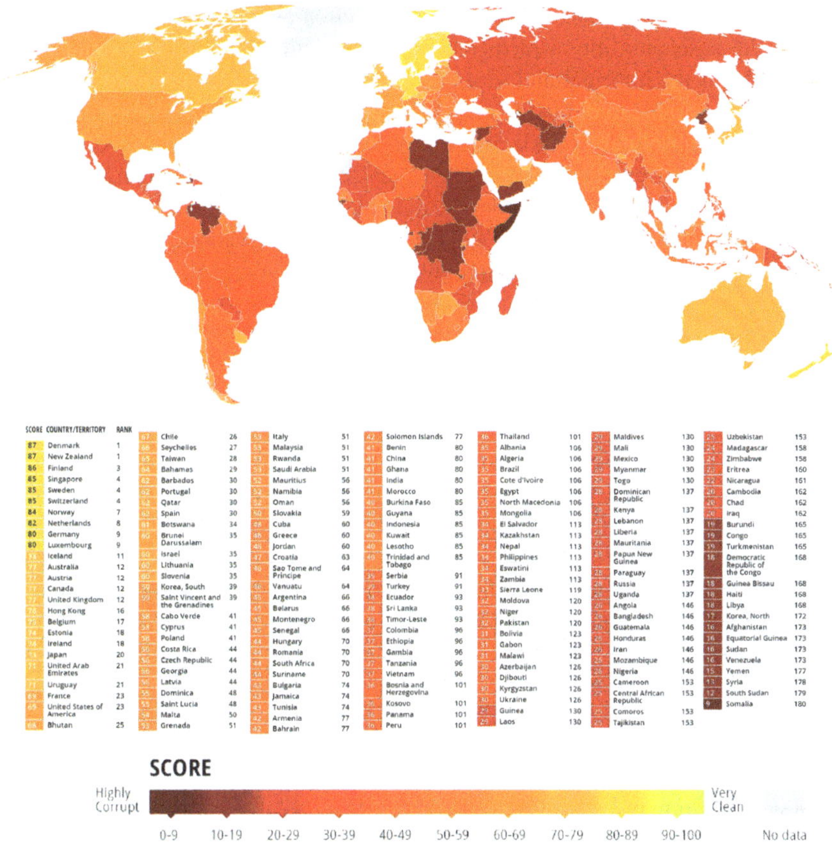

Fig. 11 The CPI 2019 around the world. *Source* [57]

Tanzanite, Platinum, Palladium, Rhodium, Iridium, Diamond, Copper involve underground mining processes, in which without a thorough investigation, such mining activities can easily increase the risk of MSIs in Africa.

Considering that, the majority of African countries produces such precious (valuable) metals (Fig. 1 and Table 2); thus implementable policies, dedicating resources, periodic auditing teams of manufacturers and retailers, interventions, an integrative framework with external institutions, among other recommendations, should be considered to eradicate MSIs. Since modern slavery is a global problem, this means both stakeholders and shareholders must work synergistically in dealing with such problems. The eradication process should not be left to the specific company, firm or organisation which deals with the specific luxury products.

MSIs are habitual practices to the extent that probably people come across it regularly. Eradicating MSIs thus need multidimensional reactions which address the range of influences—legal, cultural, social and economic—which cause human danger and facilitate abuses [30]. The African governments are encouraged to perform the following [30]. (a) To foster strong social protection as this is required to counterbalance the susceptibilities which impose people into MSIs; (b) to extend labour rights both for the informal and formal economy—where MSIs are presumably to happen—is desirable to defend labours from abuses; (c) enhanced robust migration governance is absolutely significant to protect victims and avert forced labourers because migration officers can trace a large share of MSIs; (d) to reinforce and broaden national studies and appropriate data-gathering undertakings about MSIs to lead national policy consideration; and (e) strengthen international partnerships and collaboration in tackling MSIs is vital given its cross-border and global dimensions.

6.2 Social and Practical Implications

This chapter lays significant fundamental implications for society and practice. The considered societies are the African people. The awareness concerning MSIs should be heightened within the entire supply chain of luxury products. Manufacturers as the major engine to process such products should be the frontline actors in collaboration to retailers to eradicate any sort of slavery practice within the entire supply chain. The society should be encouraged and assured protection from relevant authorities for them to disclose slavery practices. This should also be performed through training to society and other stakeholders on how they may identify the modern slavery symptoms.

This research lay a fundamental concept of how sustainable industrialisation framework in eradicating modern slavery can influence public policymakers. For example, the study suggests that the African Union should emphasise all fifty-five members to enact appropriate laws which can directly enforce companies and organisations or individuals who might engage in MSIs in manufacturing and/or retailing luxury products. Thus, Africa needs consolidative task forces that would help to fight the diverse environment which results in MSIs. Finally, by abiding on these findings, there is a contribution in influencing the public attitudes to affect the quality of life for the entire African societies. Obviously, the findings should be treated as provisional because they are based on secondary sources.

6.3 Future Research

Notwithstanding several recommendations from several researchers, nevertheless, the African model of solving such MSIs should consider the fiscal space of the country and the governing structure, specifically at each local government and sub-national level, if possible. There is a need to investigate how older people are treated in several industries, including the textile industries. For example, several older people are commonly found working in manufacturing some products. Although not necessarily being luxury products, but physically the older people, are more forced to work either due to the poverty status or employing them might be another kind of slavery. Such a study should investigate the proportion of salary paid to them in comparison to young adults in the same companies using equity theory perspectives. This should be investigated by looking for any form of exploitation for older people who work without their control. Several studies include or discuss child labour as part of modern slavery, but hardly any research investigated older people as a probable part of modern slavery forms. Therefore, future research should ponder on this influential group of society.

References

1. Akther R (2014) Redefining the meaning of Luxury goods: A conceptual paper. Beagle J Stud Res Enterprise 2:1–10
2. Al Jazeera (2018) Mapping Africa's natural resources: an overview of the continent's main natural resources. In: Al Jazeera Media Network. https://www.aljazeera.com/indepth/interactive/2016/10/mapping-africa-natural-resources-161020075811145.html. Accessed 7 May 2020
3. Amnesty International (2020) Is my phone powered by child labour? In: Amnesty International and Afrewatch. https://www.amnesty.org/en/latest/campaigns/2016/06/drc-cobalt-child-labour/. Accessed 27 Apr 2020
4. Amnesty International (2017) Time to recharge: corporate action and inaction to tackle abuses in the cobalt supply chain. London, UK
5. Anti-Slavery International (2020) What is modern slavery? In: Anti-Slavery International. https://www.antislavery.org/slavery-today/modern-slavery/. Accessed 18 Apr 2020
6. Arnegaard TJ, Davis T (2019) Mapping of modern slavery and recommendations for the Norwegian Government's development programme to combat modern slavery. Norwegian Agency for Development Cooperation (Norad), Norway
7. Ati HAA (2017) Human Smuggling and Trafficking in Eastern Sudan. Report no. 2, Chr. Michelsen Institute (CMI), Norway
8. Bales K, Hesketh O, Silverman B (2015) Modern slavery in the UK: How many victims? Significance 12:16–21
9. BBC (2019) Top tech firms sued over DR Congo cobalt mining deaths. In: BBC. https://www.bbc.co.uk/news/world-africa-50812616. Accessed 3 Mar 2020
10. Boddy-Evans A (2019) Types of Slavery in Africa and the world today. In: ThoughtCo. https://www.thoughtco.com/types-of-slavery-in-africa-44542. Accessed 22 Apr 2020
11. Bureau of International Labor Affairs (2016) List of goods produced by child labor or forced labor. United States Department of Labor, Washington, D.C.
12. Callaway A (2018) Powering down corruption tackling transparency and human rights risks from Congo's Cobalt mines to global supply chains. The Enough Project
13. Cambridge University Press (2020) Luxury goods. In: https://www.cambridge.org/. https://dictionary.cambridge.org/dictionary/english/luxury-goods. Accessed 20 Apr 2020
14. Caniato F, Caridi M, Castelli C, Golini R (2011) Supply chain management in the luxury industry: A first classification of companies and their strategies. Int J Prod Econ 133:622–633
15. Cdebaca L, Sigmon JN (2014) Combating trafficking in persons: a call to action for global health professionals. Glob Health Sci Prac 2:261–267
16. Christ KL, Rao KK, Burritt RL (2019) Accounting for modern slavery: an analysis of Australian listed company disclosures. Account Audit Account J 32:836–865
17. Cooke B (2003) The Denial of slavery in management studies. J Manage Stud 40:1895–1918
18. CPP Luxury (2019) Africa: the luxury market of the future? In: https://cpp-luxury.com/. https://cpp-luxury.com/africa-the-luxury-market-of-the-future/. Accessed 21 Apr 2020

19. Crane A (2013) Modern slavery as a management practice: Exploring the conditions and capabilities for human exploitation. Acad Manag Rev 38:49–69
20. Dahir AL (2018) Africa is now the world's epicenter of modern-day slavery. In: Quartz Africa. https://qz.com/africa/1333946/global-slavery-index-africa-has-the-highest-rate-of-modern-day-slavery-in-the-world/. Accessed 25 Feb 2019
21. Danese P, Romano P, Vinelli A (2016) Logistics and supply chain management in luxury fashion retail: empirical investigation of italian firms. a review and outlook. In: Bartezzaghi E, Cagliano R, Caniato F, Ronchi S (eds) A journey through manufacturing and supply chain strategy research. Springer, Cham, pp 169–198
22. Freedom House (2018) Freedom in the world 2018. In: Freedom House, Washington. https://freedomhouse.org/country/chad/freedom-world/2018. Accessed 27 Apr 2020
23. Gardetti MA (2020) The basic aspects of luxury and its relations with sustainability. Environmental footprints and eco-design of products and processes. Springer, Singapore, pp 17–64
24. Gardetti MA, Justo R (2017) Sustainable luxury fashion: the entrepreneurs' vision. In: Gardetti MA (ed) Sustainable management of luxury. Environmental Footprints and Eco-design of Products and Processes, Springer, Singapore, pp 347–360
25. Gold S, Trautrims A, Trodd Z (2015) Modern slavery challenges to supply chain management. Supply Chain Manag 20:485–494
26. GOV.UK (2018) UK fashion brands take action to tackle modern slavery. In: Prime Minister's Office. https://www.gov.uk/government/news/uk-fashion-brands-take-action-to-tackle-modern-slavery. Accessed 25 Feb 2019
27. Hartnady CJH (2009) South Africa's gold production and reserves. S Afr J Sci 105:328–329
28. ILO (2017a) Regional brief for Africa: 2017 global estimates of modern slavery and child labour. International Labour Organization (ILO), Geneva
29. ILO (2017b) Methodology of the global estimates of modern slavery: Forced labour and forced marriage. International Labour Organization (ILO), Geneva, Switzerland
30. ILO (2017c) Global estimates of modern slavery: Forced labour and forced marriage. International Labour Organization (ILO) and Walk Free Foundation, Geneva
31. ILO (2010) The ILO and Corporate Social Responsibility (CSR). The International Labour Organization (ILO), Geneva
32. Jesperson S (2019) Conflict and migration: From consensual movement to exploitation. Stab Int J Secur Develop 8:1–13
33. Krueger B (2006) Solving slavery in Sudan: solving a social problem through nation building. Office of Honors Programs & Academic Scholarships, Texas A&M University
34. Lalwani SK, Nunes B, Chicksand D, Boojihawon DK (Roshan) (2018) Benchmarking self-declared social sustainability initiatives in cocoa sourcing. Benchmark Int J 25:3986–4008
35. Larsen JJ, Diego-Rossell P, Bermudez LG et al (2019) Modelling modern slavery risk. United Nations University, New York
36. Li J, Leonas KK (2019) Trends of sustainable development among luxury industry. In: Gardetti MA, Muthu SS (eds) Sustainable luxury. Environmental Footprints and Eco-design of Products and Processes, Springer, Singapore, pp 107–126
37. Lodestar Marketing Research (2015) The luxury fashion industry in South Africa. Industry report, Switzerland Global Enterprise, Switzerland
38. Macias-Konstantopoulos W (2016) Human trafficking: The role of medicine in interrupting the cycle of abuse and violence. Ann Intern Med 165:582–588
39. Marshall D, McCarthy L, McGrath P, Harrigan F (2016) What's your strategy for supply chain disclosure? MIT Sloan Manag Rev 57:37–46
40. Muralikrishna IV, Manickam V (2017) ISO 9000, 14000 Series, and OHSAS 18001. In: Environmental management: science and engineering for industry. Elsevier Inc, pp 177–208
41. New SJ (2015) Modern slavery and the supply chain: the limits of corporate social responsibility? Supply Chain Manag 20:697–707
42. O'Driscoll D (2017) Overview of child labour in the artisanal and small-scale mining sector in Asia and Africa. K4D Helpdesk Report, the UK Department for International Development (DFID), London

43. Oxford University Press (2019) Luxury. In: Lexico.com. https://www.lexico.com/definition/luxury. Accessed 20 Apr 2020
44. Prendergast G, Wong C (2003) Parental influence on the purchase of luxury brands of infant apparel: An exploratory study in Hong Kong. J Consum Mark 20:157–169
45. SAnews.gov.za (2019) Chinese national nabbed for exploitation, human trafficking. In: Government Communication and Information System, South Africa. https://www.sanews.gov.za/south-africa/chinese-national-nabbed-exploitation-human-trafficking. Accessed 23 Apr 2020
46. Serdari T (2016) Experiments in suchness: hiroshi sugimoto's silk shiki for hermès. In: Armitage J, Roberts J (eds) Critical luxury studies: art, design, media. Edinburgh University Press, Edinburgh
47. Soundarya S (2015) Corporate social responsibility: a contemporary approach towards sustainable development. IOSR J Bus Manag:40–43
48. Sovacool BK (2019) The precarious political economy of cobalt: Balancing prosperity, poverty, and brutality in artisanal and industrial mining in the Democratic Republic of the Congo. Extract Ind Soc 6:915–939
49. Steele S (2013) Human trafficking, labor brokering, and mining in Southern Africa: Responding to a decentralized and hidden public health disaster. Int J Health Serv 43:665–680
50. Stevenson M, Cole R (2018) Modern slavery in supply chains: a secondary data analysis of detection, remediation and disclosure. Supply Chain Manag Int J 12:81–99
51. Stringer C, Michailova S (2018) Why modern slavery thrives in multinational corporations' global value chains. Multinat Bus Rev 26:194–206
52. Stronger Together (2018) Prominent businesses join programme to address forced labour in the fruit and wine producing industries in South Africa. In: Stronger Together. https://www.stronger2gether.org/prominent-businesses-join-programme-to-address-forced-labour-in-the-fruit-and-wine-producing-industries-in-south-africa/. Accessed 23 Apr 2020
53. Such E, Laurent C, Salway S (2017) Modern slavery and public health. https://www.gov.uk/government/publications/modern-slavery-and-public-health/modern-slavery-and-public-health. Accessed 3 Mar 2020
54. Taifa IWR, Hayes SG, Stalker ID (2020) Development of the critical success decision criteria for an equitable order sharing in an extended enterprise. TQM J 32: 1715–1742. https://doi.org/10.1108/tqm-05-2019-0138
55. Taifa IWR, Hayes SG, Stalker ID (2020b) Computer modelling and simulation of an equitable order distribution in manufacturing through the Industry 4.0 framework. In: 2nd International conference on electrical, communication and computer engineering (ICECCE), 12–13 June 2020. IEEE, Istanbul, Turkey, pp 1–6. https://doi.org/10.1109/ICECCE49384.2020.9179275
56. Taifa IWR, Hayes SG, Stalker ID (2020c) Towards a digital revolution in the UK apparel manufacturing: an Industry 4.0 perspective. In: Bartolo P, Silva F, Jaradat S, Bartolo H (eds) Industry 4.0–Shaping the Future of the Digital World, 2nd International Conference on Sustainable Smart Manufacturing (S2M 2019), Manchester, 1st edn. Taylor & Francis, London, pp 3–8
57. Transparency International (2020) Corruption Perceptions Index 2019. https://images.transparencycdn.org/images/2019_CPI_Report_EN_200331_141425.pdf. Accessed 2 Oct 2020
58. UK Hm Government (2019) UK annual report on modern slavery. Modern Slavery Unit, Home Office, London, UK
59. UN DESA (2019) World population prospects 2019: data booklet. Department of Economic and Social Affairs (DESA) Population Division, United Nations, New York, USA
60. United Nations (2020) Sustainable development goals. In: United Nations Department of Public Information. https://sustainabledevelopment.un.org/?menu=1300. Accessed 28 Apr 2020
61. United States Department of State (2017) Madagascar 2017 Human Rights Report. USA
62. van der Wal S (2019) Child labour in Madagascar's Mica sector: Impact of the mica supply chain on children's rights from the Malagasy mines to the international product line. The Centre for Research on Multinational Corporations (SOMO), Amsterdam

63. Vigneron F, Johnson LW (2004) Measuring perceptions of brand luxury. J Brand Manag 11:484–506
64. Virtanen A (2020) Corporate social responsibility reporting in the European apparel industry: the challenge of pursuing greater accountability and transparency. Master's Thesis, School of Business and Management, Lappeenranta-Lahti University of Technology (LUT)
65. Walk Free Foundation (2018) The Global Slavery Index 2018. The Minderoo Foundation Pty Ltd (MFPL), Nedlands Western Australia
66. World Economic Forum (2018) Driving the sustainability of production systems with fourth industrial revolution innovation. Geneva
67. World Gold Council (2018) Slavery and human trafficking statement of the world gold council for the financial year 2018. World Gold Council, UK

Lab-Grown Diamond–The Shape of Tomorrow's Jewelry

Anil Kumar Bagathi, Carmelo Balagtas, Sai Vijay Kumar Boppana, Ivan Coste-Manière, Florent Vincent, François Le Troquer, and Gérard Boyer

Abstract Lab-grown diamond: such a mysterious description embodied in the association of one of the most famous gemstones on earth with the technical know-how that scientists developed through years of research. Mostly known for being one of the precious gifts to offer for a significant event such as engagement and wedding, diamonds continued to spread its admiration for the dazzling effect which

---

A. K. Bagathi · C. Balagtas · S. V. K. Boppana · I. Coste-Manière (✉) · F. Vincent · F. Le Troquer · G. Boyer
SKEMA Business School, 06902 Sophia-Antipolis, France
e-mail: ivan.costemaniere@skema.edu

A. K. Bagathi
e-mail: anilkumar.bagathi@skema.edu

C. Balagtas
e-mail: carmelo.balagtas@skema.edu

S. V. K. Boppana
e-mail: saivijaykumar.boppana@skema.edu

F. Vincent
e-mail: florent.vincent@skema.edu

F. Le Troquer
e-mail: fletroquer@hotmail.com

G. Boyer
e-mail: gerard.boyer@univ-amu.fr

G. Boyer
Université Marseille Aix, 13007 Marseille, France

I. Coste-Maniere and M. Á. Gardetti (eds.), *Sustainable Luxury and Jewelry*, Environmental Footprints and Eco-design of Products and Processes,
https://doi.org/10.1007/978-981-16-2454-4_11

convinced more than one couple. Since the luxury industry is migrating towards a more conscious way of manufacturing products and consumers are thriving towards ethical consumption and bold products, lab-grown diamond is gaining in popularity against the bleak stories of the mined ones. Keeping this epigram in spotlight, this article provides the complete view of how the industry is going to adapt to this new revolution. Lab-grown diamonds are diamonds that have been processed in laboratories thanks to almost 70 years of development. To test the idea that this new alternative is a legitimate opportunity to answer the need of the current changing jewelry market, this research tried to observe the real potential of this man-made gemstone through a deep understanding of its hallmark, perception of the market and concrete sustainable solutions available. As a result, it is observed that a certain interest in finding a new alternative with the same attributes to reassure all parties. What defines a gemstone is simply the association between its chemical composition and crystal structure. To this fact it created two sided views for which there is no right or wrong position but simply a question of brand positioning. New generations of consumers increasingly recognized the importance of ethical consumption of a luxury product. Bitter lines in the news, child labor and conflicts such as "Blood diamonds" created an uncertain environment in which the new generation Gen Z, nowadays does not feel comfortable to buy unconsciously. Again, pros and cons are rising over laboratory grown diamonds concerning its real repercussion on the environment. Facts are here to prove that this alternative is a relevant approach to the current luxury industry: providing high clarity and colorless gemstone, controlled production quantities, improved traceability issues, changes in price range. Luxury brands should keep educating its audience by disclosing clearly the differences between mined and lab-grown diamonds and use this increasing awareness to promote new responsible jewelry collection that relate to today's society needs.

Keywords Lab grown · Diamond · Sustainability · Man made · Environmental impact · Luxury · Ethical jewelry · Synthetic diamond

1 Introduction

Traces of the greatest inventions of all time, would recall judgements over the idea presented than a celebration of the creative processes that were responsible for what people enjoy today.

Diamond has always been considered as a staple commodity as it can last forever. In recent times, there have been controversies on what truly makes up a diamond. Words have been said and sides have been clarified. When the world has known a particular way of producing a gem, it may not be easy to persuade the consumers to trust a new one. There are many emotions attached to the creative process of a fine commodity that has been loved today and consequently, it will take significant efforts to disrupt that.

Yet it is not difficult to recollect some great forms of innovations that were once frowned upon and have only proven otherwise. "The iPhone is nothing more than a luxury bauble that will appeal to a few gadget freaks."—Matthew Lynn, Bloomberg, 2007. "She doesn't wear fur because she is cold, she wears fur because she is glamorous."—Ken Downing, Neiman Marcus VP. "The horse is here to stay but the automobile is only a novelty. A fad."—A bank President's advice to not to invest in Ford Motor Co, 1903. It is rather easy to place a bad judgement on upcoming idea, than at least be curious about it. But an idea that holds relevance and purpose would not be easy to ignore. These commodities, supplemented by technology, are still around to date and have been game changers since the birth of its conception. And just like any of the greatest inventions found in the world, it would only take an excellent value proposition to re-state a story in the Jewelry sector—more particularly the Diamond creating process.

The research will begin with a comprehensive technical discussion from conception to implementation. The first section will contain components for comparison of both natural and man-made diamonds. Slowly, the transition will be made to understand how marketable this is in today's consumer's interests. Diamond enthusiasts will begin to understand the relevance of this commodity as the trend spreads popularity to the rest of world. And finally, the last section shall give insights on the sustainability measures that both types of diamonds possess both on the conservation of the environment and the protection of its people. The intent of this research is to add value to existing researches made, link those who are interested to those who already identify themselves as enthusiasts and later hopefully create a bond to continue this story-telling that started all the way from the 17th century.

Developments are pro-actively done to re-consider what has been known for a long time and accommodate a new. And while it will only take time for the world to be convinced in believing in the integrity of an alternative—fine commodity, may this research spark curiosity to discover this new trend that is certainly taking place in front of our eyes called the Lab Grown Diamond.

2 History and Techniques of Lab Grown Diamond

2.1 The Birth of Lab-Grown Diamonds

Hearing about laboratory grown, the imagination could lead to clones and huge chemical explosion. It has the particularity to have a certain modern and scientific approach. To this fact, mixing the word laboratory-grown with one of the hardest and oldest stone on earth, diamond, could create some confusion. Lab-grown diamond is the culmination of technological development in terms of recreation of natural material. Most people think that this innovative technology is very new but, the story of lab-grown diamond started a long time ago, more than a century old. Since the discovery of the chemical composition of diamond by the French scientist Antoine Lavoisier in 1772, many chemists and powerful companies attempted to create man-made diamond by heating charcoal without positive results. The conditions of diamond crystallization are very hard to set up because it requires extreme temperature and pressure. Fortunately, the industrial revolution has set in motion and developed technological advanced instruments in all fields thus providing new tools for scientists to do research.

The history of laboratory grown diamond started back in the 50's when the inventor named William. G. Eversole makes the first very small Chemical Vapor deposition (CVD) lab-grown diamond within the Union Carbide Corporation. The

Swedish company called Allmänna Svenska Elektriska Aktiebolaget (English translation: General Swedish Electrical Limited Company; Abbreviation: ASEA) made immediately, approximately one year later, the first High pressure High temperature (HPHT) lab-grown diamond. The hype and the awareness of lab-grown diamond industry initiated when the American group General Electrics (GE) produced in 1954 their HPHT lab-grown diamond for industrial purposes and inaugurated their discovery on media and newspapers. Following this, General Electrics continued to invest in the development of this technology to give after 13 years of research, in 1970, the first quality synthetic diamonds. Even if for the moment the production runs were colossal, it remained an exceptional innovation in the history of chemistry. A new market in synthetic stone was therefore emerging directly for modern jewelry, which means that other countries and companies will consider the subject later.

In the 1990s, Russia successfully developed less expensive HPHT production methods, thus providing a financially affordable synthetic diamond to enter the gemstone market. In the same period, the American diamond giants GE and DeBeers were initiating experiences by inserting new components to create variations of synthetic diamonds: colored lab-grown diamond.

Then in 2000s, American companies such as Gemesis, Appolo and Chatham took advantage of technological advancements to refine recipes, reduce production costs and greatly improve the quality and size of the synthetic diamonds produced. The latest progresses have made possible to create diamonds of perfect clarity (D) with very little inclusion (VS) which represents an extremely rare quality in the natural state of mine diamonds. A Russian company by the name of New Diamond Tech (NDT) has held a size record since 2018 with an HPHT diamond weighing 103.5 carats, while DeBeers launched in May 2018 "Lightbox", the commercial platform for lab grown diamond with a retail price of 800 dollars per carat. They set the standard and start to design how lab grown diamonds should enter the market in the digital era. Today, the first French lab-grown maker named "Diam Concept" performed CVD lab-grown diamond from 0.5 to 2 cts. Diam Concept is working in collaboration with the house of jewelry Courbet who is entirely dedicated to provide a fully ethical supply chain and jewelry products.

After having drawn a timeline on the evolution made for the lab-grown diamond industry, this research will pursue on the understanding the real definition of lab-grown diamond and how it differs from the mined one.

2.2 *Unbiased Description*

The first concern that emerged from lab-grown diamond is its interpretation. People who were not aware of this new alternative might find confusing to evaluate its true nature and market value. A diamond, whether formed in a kimberlite cone of an ancient prehistoric volcano or in the ultra-sophisticated laboratories of scientists is still a diamond. The institute responsible for the regulation and laws of business trades of new materials in the United States, the Federal Trade Center (FTC) has indeed changed the wording of the description of diamond several times in order to satisfy all parties.

Since 2018, the definition of diamond evolved by deleting the terms "natural" from its description. According to the FTC, a diamond is "*a mineral consisting essentially of pure carbon crystallized in the isometric system*"[1] including both lab-grown and mined gems.

Most importantly, the improvement and clarifications were made around the type of disclosure that must be associated to the lab-grown diamonds. The terms recommended by the FTC to disclose properly and differentiate a lab-grown diamond from a mined one are the following: "*laboratory-grown,*" "*laboratory-created,*" "*[manufacturer name]-created,*" "*imitation*" *or* "*simulated*".[2]

Indeed, this applies for the American market, but differences might appear in other countries. Taking the example of the French regulations, the term "cultured diamond" is accepted in the US under certain conditions whereas it is totally forbidden in France.

Meaningfully, it is crucial to understand that the legislation around synthetic stones is updated more efficiently in some countries where they have been established for a longer time such as the United States while in France or in Europe the market is still young. Foreign Trade Centre (FTC) latest updates for lab-grown gemstone were made in 2018 whereas the French law were lately modified in 2002. To develop an ethical and responsible trade of laboratory stones, it is important to adjust the regulations as well as the control of the correct in-store presentation so that they can be exploited at their fair value (Fig. 1).

2.3 Making Process

The physical properties of a diamond whether naturally mined or created in a laboratory are identical. It possesses the same hardness ranked at the first position on the Moh's Scale with an index of 10/10. Its crystal structure is similar with an isometric face centered cubic structure composed of tetrahedra carbon atoms liaisons. Their chemical compositions are likewise identical with 99.95% of carbon with 0.05% of remaining particles.

To achieve this result, scientists have developed various machineries that are inspired by the conditions in which diamonds were created in the natural environment. Producing such pure crystallization that respects the same crystal structure requires a lot of energy, which is particularly challenging to concentrate. There are two major techniques nowadays to create diamonds.

The first method of growth is manufactured by High Pressure High Temperature (HPHT). This approach replicates what happens in the earth's crust during the creation of diamonds deep in the rock mantle. It is run with six hammer-presses positioned in three dimensions which will exert a pressure of about 5 to 6 GPA

[1] Graff [20].

[2] Fair and Frisby [17].

Fig. 1 ©GROMIK Thierry. Rough Lab-grown surrounded by fine Jewelry selection from Courbet

on all sides of the cubes of a diamond seed base (itself natural or synthetic). The combustion chamber which contains the seed must withstand this high pressure and be heated to a temperature of 1300–1600 °C to add the mineral powder which will allow the diamonds to form on the base. This process, being the most common, can produce large laboratory-grown diamonds with high clarity. The color quality may vary depending on purity of the diamond with whitish to greyish hue and if additional trace elements were added during the growth process. For instance, it can provide blue-colored diamonds with boron or yellowish one with nitrogen. The lead time is estimated around days and weeks following the type of diamond expected.

The second method is performed using Chemical Vapor Deposition (CVD). This involved using a main base of gas like methane to generate a high temperature plasma from 700 to 1000 °C. This method necessitates a capsule with a much lower pressure than for the HPHT method. Something around 2500 pa to 26000 pa of pressure but the CVD method requires advanced technology on the use of microwave radiation to control and diffuse a beam of 2.45 Ghz. It will be followed by a coating that will appear slowly at the diamond seed surface adding thickness and creating the diamond layers after layers. The CVD method is very sophisticated which makes it an even more difficult tool to calibrate but allowing to obtain diamonds of rare quality. CVD-grown diamonds are much expensive to create but they have the purest clarity and colors like the finest mined diamonds on earth that are well researched by the high-end jewelry industry. At this point, high purity colorless diamonds are the most coveted ones. Moreover, depending on the wishes and needs of the CVD lab-grown maker, it is possible to add certain substances in order to vary the color slightly and give white, pinkish to reddish, or brownish colored diamond tones. The quality of the mother seed is essential for a perfect development of crystallization and to avoid the birth of cracks or unwanted particles.

HPHT and CVD are the two main methods used to cultivate lab diamonds. Depending on the type of diamond desired considering the size, purity and utility, the choice of method should be made wisely. As for fine jewelry, HPHT-colored diamonds have been chosen but over the years, there is a shift in focus towards a choice of CVD process. Each method has its advantages and disadvantages. The HPHT method generally promotes the creation of larger diamonds with faint colors while the CVD method will form smaller but very pure and colorless diamonds.

Furthermore, patents already reserved and available in China and Germany are provided to reveal some of the latest technical progress in the manufacturing process. According to the CNRS (the french National Center of Scientific Research), many innovations cropped up last year aiming to reduce the carbon footprint from diamond manufacturing, improving the quality of lab-grown diamonds produced and finding new cost effective method of production.

The first patent is proposing a new body structure and combustion chamber. The body structure is made out of "molded tubular structure of graphite material" that would be heated into a bearing structure. Improvements brought through this method are using less energy as the temperature to synthetize in the combustion chamber should be lower than traditional method, enhancing in the same time the quality of the diamond.[3]

The second innovation that will increase the gem quality and consistency are a carved tube to prepare and present the powder more efficiently into the combustion chamber and a manufactured block that can be inserted into the combustion chamber. The purpose of these advancements is mostly oriented around the intent to raise the gemstone's production accuracy.[4]

Another engaging device was also created to upgrade the current machine that produce artificial diamonds. This patent consists in a mechanism that can remove automatically the grown diamonds produced without worrying about inserting traces elements which is the main priorities of lab grown diamonds manufacturer. The quest of quality began inside the forge in a pocket that should be totally airtight and this device will allow you to extract the grown diamonds during an on-going process. This therefore has a direct positive impact on the production capacity and prevent unwanted impurities.[5]

The last method this research wanted to highlight is about an intelligent capsule containing the ingredients necessary for the creation of diamonds. This capsule could then be inserted into a suitable machine. This principle facilitates the setting up of production as well as making it possible to give precise quantities of previously

[3] Li et al. [24].

[4] Cai et al. [12].

[5] Zhang [39].

prepared ingredient. This ingenious configuration is without doubt one of the most promising technological advances.[6]

2.4 *Certification or Certain Fiction?*

The definition of lab-grown diamonds is clear, it is 100% diamond. Yet, the only thing that varies is the method of making, which try to reproduce the physical reaction that happened millions of years ago under the earth's crust. The issue is the one following: Changing the source and pathway of the stone making process will affect its storytelling which most of the High jewelry Houses are focusing for their marketing campaign.

To this fact, the question remains in the ability to identify properly which stone is man-made and which one was extracted from a large mine. The solution is among all the identification process which is probably the main problematics of jewelers and stone dealers nowadays. After dealing with counterfeit and simulant for which this research will go deeper in the next texts, stone dealers must make sure that a specific diamond is whether mined or lab grown. Thus, they will be able to properly disclose it to the concerned customer. To identify the nature of a diamond fabrication, even for an expert with a trained eye is very difficult. The first and maybe the prior information that must be analyzed are the inclusions. For example, for colored lab-grown gemstones such as sapphire or other, a diagnosed precise inclusion patterns are obtained due to the specific method of making. The earliest method was the flame fusion conceived in 1902, also entitled "Verneuil" method, in which we can observe inclusions such as curved lines and bubbles.

However, having a unique way of making, lab-grown diamonds are thus even more difficult to discern tiny characteristics. It may happen that some particles of mineral powders such as sulfide, iron, nickel, cobalt inclusion appears after manufacture. These can leave tiny traces invisible to the naked eye like black grooves and therefore it is possible to identify a laboratory diamond.

Some tools and tests can help gemologist to provide clues for stone identification. Among them, there are the Magnetism test, the Infrared Spectroscopy test, or even the Electrical conductivity measurement test. Even if these alternatives do not provide concrete solutions, they contribute to gather a panel of information necessary for gemological experts to support their decision-making. The in-depth analysis of inclusions will always have the final word on the diamond appraisal.

As mentioned previously, being one of the most precious stones on earth, diamonds are a highly coveted resource that will stir up the reflections of counterfeiters. It is the disclosure that will differentiate a simulant from a counterfeit. In fact, a simulant is defined by a stone which reproduces the characteristics of another stone by its brilliance, sparkling, color, and purity. A simulant therefore has no similarities

[6] Matjuschkin and Woodland [28].

neither in terms of crystal structure nor chemical composition, which will significantly influence its hardness and toughness. A fake diamond is in a way a simulant that was disclosed as a real one.

To counter these criminal practices, certification appears to be the best solution to ensure the correct identification of the stone by a renown gemological institute. Among the most popular, there are the Gemological Institute of America (GIA), the French laboratory of gemology (LFG), and the Belgium HRD Antwerp that are specialist in diamond certification. Since October 2020, GIA has even offered a fully digital option to access their diamond certificates, allowing flexibility and practical use through smartphone app. Guaranteeing the veracity of the identity of a stone can be expensive and so it is only applied for exceptional stones.

However, certification is also a subject to be conscious about. A grey market is existing for fake certificates as any other credential. In addition, even if the eye and the knowledge of an expert are the best choice, the buyer must not forget that it remains subjective. The analysis will always be affected by the country in which it is established, with its regulations, habits and its cultural background that will influence the investigation of the stone. There will be no miracle on identification if there is no clear clarification on the traceability of the stones. For all scarce resources and sources of income in poor country, the financial side will always be a priority. Fortunately, the hope for a more ethical future is appearing thanks to modern technologies which bring new tools to improve traceability. The major institute GIA remains far the leader with developed identification capacities and strong gemological databases on which they can rely on (Fig. 2).

Fig. 2 ©GROMIK Thierry. Lab-grown diamond identification with a magnifier

3 Marketing and Feasibility

3.1 New Story in the Contemporary Trend

In Luxury two components remain to be overused as if it's the only way through success. Heritage defined as—entity that is handed down from the past as a tradition, and a story either true or fictitious, designed to interest, amuse, or instruct the reader, tale. These two components out of many sustain a dialogue between a brand and a customer. Heritage and Story therefore introduce a brand for awareness, entice audience for interest, and sustain that dialogue for loyalty.

DeBeers, the diamond giant, has created the first luxury campaign exclusively focused on storytelling and diamond development. In 1999, De Beers' campaign "A Diamond is Forever" was the most memorable slogan of the twentieth century. But the campaign, which proposed the idea that no marriage would be complete without a diamond ring, wasn't just riding on the aid of an existing industry. De Beers built the industry. It presented the idea that a diamond ring was a necessary luxury. According to the New York Times, N.W. Ayer's game plan was to "*create a situation where almost every person pledging marriage feels compelled to acquire a diamond engagement ring.*"[7] Yet the world of luxury may have magnified a different take on heritage and story. The industry has now paid attention to this emerging trend on 'Conscious Luxury' through the man-made diamonds. As lab grown diamonds become more famous, the normal rebranding effort does not seem to work for mined diamonds.

Indeed, the global demand for synthetic diamonds, from jewelry, industrial and medical devices is expected to be US$28.6 billion by 2023, up from about US$15 billion in 2014, according to Crystal Market Research.[8] Successively, another report stated that every year the production of lab-grown diamonds is growing by 15% and in the same source Forbes article reported, that with the rapid increase in the production of synthetic diamonds, the $80 billion mined diamond industry will soon be disrupted.

The tide is turning around. The diamond industry faces problems of oversupply and a downturn in demand across the globe. In order to cope with the pressure from the wholesale market, De Beers shares that they will use money to boost retail efficiency in China and US. Unlike the United States, which has been buying diamonds for more than half a century, China is still a relatively new market for diamonds yet one with great purchasing potential.

As the consumption of diamonds has increased in such a short time, stakeholders are encouraged to continue to bet on China's diamond market. In 2017, according to De Beers Group's, "Diamond Acquisition Report China"[9] love-related gifting dominates China's diamond market, accounting for 58% of diamond acquisitions.

---

[7] Kolowich Cox [23].

[8] Wilkinson [35].

[9] Wu [37].

True until today, ‘Couple rings’—a set of matching designs—are common among prospective Chinese spouses. In fact, according to the Diamond Insight Report 2019 published the most famous form is a platinum ring with diamonds, accounting for 74% of the couple rings acquired in China. Also, the proportion of Chinese brides receiving diamonds has moved from 0 to 47% in less than 30 years, suggesting that diamonds have been well received by Chinese consumers. Another market worth visiting is in the capital of Luxury itself. Taking France as an example which is the 3rd biggest luxury market, it has been reported by the study ExJewel published by the author Stephane Boghossian that in France, it has sold more lab-grown diamonds than natural diamond Jewelry.[10] Two components were identified that have caused to such selling, including the lower price tag—which is around 30 to 40 % cheaper than a natural diamond—and the change in jewelry purchasing behavior.

The shorter supply chain is part of the explanation for the reduced price. Mined diamonds have an intricate and lengthy channel, passing through a variety of industries and companies to get from a mine to a jeweler. Laboratory diamonds avoid the need for a long supply chain, reducing the costs of middlemen[11]. While not cheap, lab-grown diamonds provide a less expensive alternative to traditionally mined gemstones. Consumers can buy the same product—a dazzling diamond—at a reduced price. The other component highlights the brilliance that fine piece carries, which has been accredited by GIA to be at par with the natural diamond, more specifically components such as purity, density, and color. Supplementing the entire idea that regardless how this commodity is made—laboratory or natural—it is technically considered as a real Diamond (Fig. 3).

3.2 *Selling a Jewel of Values*

Several elements are considered before purchase by consumers which include quality, brand influence and price positioning. Lab grown diamond is more favorable to consumers given its significant price difference compared to the naturally mined diamond. Understanding that different companies have different pricing approaches in place, it is worth noting the different type of strategies identified for each of the Lab grown diamond producers as cited below:

- **DeBeer**: One of the biggest producers of diamonds came into the new revolution of lab grown diamonds with the price positionings of 200$ and 800$ respectively for ¼ carat and 1 carat.
- **Courbet**: This company is selling the same price as how diamond jewelry is sold. They have become huge sellers of lab grown diamond jewelry in France.
- **Brilliant Earth**: One of the biggest companies from USA that sells LGD’s. It has priced their diamonds of 1.25 times than the natural diamond.

[10] Boghossian [11].

[11] Little Switzerland [26].

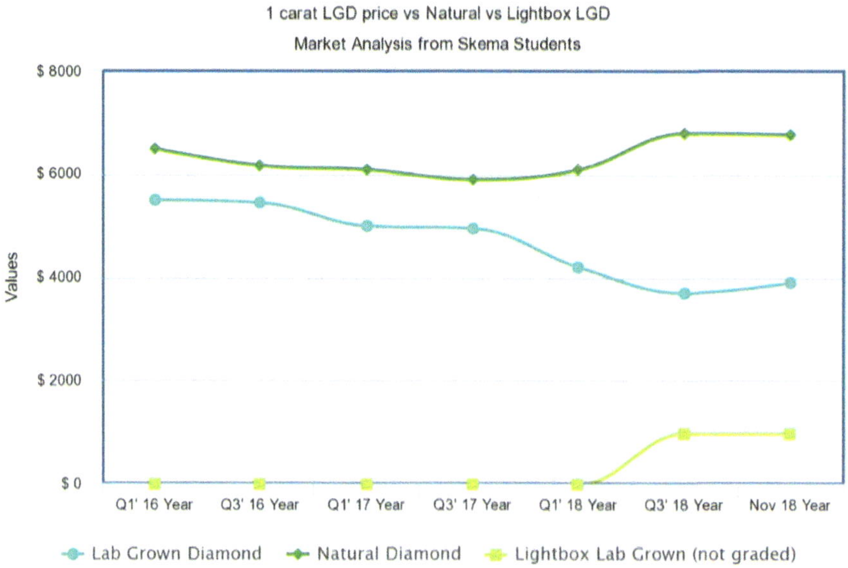

Fig. 3 Spline Chart designed by the Skema Students according to the information retrieved from the Paul Zimnisky's analysis. ExJewel. November 24, 2018

- **Diamond Foundry**: Also from the United States and selling their diamonds at 1/3 of price of natural diamond.

According to an article published beginning of 2020 by Ex-Jewel,[12] based on Google's global search and press coverage, the list of lab-Grown diamond jewelry brands, start-ups and projects in the ethical and sustainable jewelry industry, is the following. These relatively famous players in this emerging business proposition are enlisted,

1. Diamond Foundry
2. Brilliant Earth
3. Courbet
4. Lark & Berry
5. Lightbox (DeBeer's)
6. Swarovski
7. Innocent Stone
8. Kimaii

Ranked at the third position, Courbet is a reputable brand who is setting the stage for French Lab Grown Diamond in the epicenter of Luxury. Courbet opened its doors in May 2018 with a store on Place Vendôme in Paris, possibly the most prestigious jeweler's address in the country, counting luxury reliable Cartier and Van Cleef & Arpels as neighbors. Fine tuning its presence in the luxury sector more

[12] Boghossian [11].

particularly in the field of Diamonds, Courbet founders Manuel Mallen and Marie-Ann Wachtmeister forecast with the new strategy in place.[13] Courbet will reach a turnover of 50 million euros in the next years to come. This backs up both the demand for a more 'Conscious Luxury' and the use of the latest technology to manufacture an alternative to the mined diamonds.

Known to be the "Jeweler of Values", owner and founder, Mr. Mallen shares, "We are proud to be different". The brand shakes the discussions by proving doubts with results that showcases offerings of magnificent jewelry with laboratory grown diamonds produced in France. In fact, a recent collection named "Pont des arts" showcased French Lab Diamonds from the south of France, while the other existing collections from Courbet sourced elsewhere following stricter regulation protocols.

> Lab-grown diamonds are at present not available in your local jewelry, but more than likely it won't be long to get in your hands

3.3 Sustaining a Conversation on Conscious Luxury

Conversations continue to unravel how this new trend can be sustained. A report published by Bain & Co. describes the three main disruptors in the mining-diamond market: digital space, lab-grown diamonds, and increasing customer appetite for environmental and social responsibility of which 'Conscious Luxury' takes front and center. The Bain & Co. study is pegging the growth of the lab-grown diamond market between 15 and 20% in 2019 and given that conventional jewelers have only recently begun to sell lab-grown jewelry. Their market share in the jewelry industry will expand and grow rapidly.

The truth today is that online experience has become inseparable from the customer's journey. The case of luxury brands is no exception. Bain & Co. 2020 report estimates that online luxury purchases will be more than tripled by 2025, with almost a quarter of personal luxury items being sold online.[14] Luxury services have followed that trend. A few years ago, it was impossible to think that people could visit live diamond stores via an application (Virtual Diamond Boutique, Blue Nile), purchase antique jewelry online from anyone willing to sell (eBay, Etsy, …), deliver it on the same day (Amazon, Farfetch) and even increase Google search.

Indeed, there is a new wave of consumers and it is imperative to mention that the new generation composed of those who were born from the years 1997–2012, the "Gen-Z" will be the new frontiers of tomorrow's luxury market. This generation along with the—Millennial will be the new consumer base to focus on. According to the research from Bain and Company partner and Frederica Levato named "Eight Themes that are rewriting the Future of Luxury Goods"[15] and published last February 2020, it was mentioned "*Millennial customers (also known as Generation Y, born*

[13] Johnson [3].

[14] Linde [25].

[15] Linde [25].

between 1980 and 1995) have been steady buyers of luxury. They account for 35% of consumption in 2019 and by 2025 could make up for 45 percent of the market. But it's the even-younger Generation Z that is poised to reshape the industry: by 2035 they could make up 40 percent of luxury buyers and they display behaviors that distinguish them from other generations." This is important to bring to the discussions as not only do these young consumers come in big numbers, but they come also with a higher esteem for the environment. Young activists showing at the United Nations and the increase media coverage and political movements like "Extinction Rebellion" are all ensuring that sustainability and climate actions are far from a trend. It is a necessity for humanity's survival. With the new consumer base leading on these initiatives, a certain level of attention that has urged brands to take action. And thus, this new "Conscious Luxury" only make sense as it is highly driven by the demand of the new frontiers of this industry.

Both Gen-Z and the Millennials' expectations from brands to the entire industry are significantly different from their predecessors. Such trend with ethically made stones at the forefront disrupt the traditional diamond industry, where this new consumer base and a more educated older generation of jewelry buyers would back up, preferring modern accessories that would reflect their spirit. It can be argued at one point that the old-fashioned jewelry may not exactly reflect the millennial's and even younger generations personal and moral values.

While Covid-19's damages may be too early to speak about, its impact has definitely highlighted what changes needed to arise and to be acted upon immediately. Time and again, the luxury industry have proven its capability to take risk and lead innovation and a Lab Grown Diamond is its latest addition to its roster. With the advent of the digital era and in line of the pandemic, certainly some of the foundational aspect of the luxury industry were shaken. Not only supply chains of these brands but the entire business will need to prepare accordingly in order to face this correctly and get it right. The natural diamond extraction method is not as easy as it seems and there are vast number of approaches in operation to consider. Reinvention of product development and hyperpolarization in performance, supplements the idea that Lab Grown Diamond is worth consumer's time and money.

The lab-grown diamond is one of the most exceptional choices where most diamond companies would look at this option for these reasons even after the pandemic is over. Sufficient to say, that it is indeed here to stay.

4 Sustainability

4.1 Why Buy a Lab-Grown Diamond?

The sparkling diamond may steal all the senses, but it may also hold a bleak, bloody secret. For several consumers, a symbol of love or gratitude, is what diamonds are for. Supplementing the idea that Love is eternal, there is no better stone than the most

Fig. 4 Man-made diamonds (2019)

enduring in nature to reflect that. Nothing tarnishes the caring act of conflict, ecological destruction, violations of human rights, and unethical processing. Unfortunately, even though doing a lot of work to guarantee that the mined diamond the consumer purchases is 'legal,' most brands just do not have the transparency to back up their statements. At the end of the day, in this long and complicated supply chain process the consumer decides or gets convinced based on the truth of the end retailer. Mindful consumers (especially Millennial's and Gen Z) are looking for the laboratory to reap the benefits of all that shimmers without costing the Planet (Fig. 4).

4.2 *The Reality of Lab-Grown Diamond*

4.2.1 Ethical

As the average consumer gets more educated about the real essence of diamond production, the ideal ethical solution is seen through the lab grown diamonds. These man-made gems have grown in popularity in recent years. In fact, according to a study by The Vendeur, laboratory-grown diamond demand has risen from 325,000 carats in 2013/14 to 4,000,000 carats in 2016/17[16] and is expected to increase more if the industry sustains the expected growth rate.

"According to a study by Princeton University the carbon footprint of a lab grown is 18-22% of a mined diamond."[17]—Alan Frampton, Jewlery director at Cred Jewelery

But the main question remains, are consumers really missing out on luxury and romance when they choose a lab grown diamond over a mined one? Powerful diamond brands have over the years told us to carry the ultimate status in rarity, glamour and sophistication. "A Diamond is forever", was a marketing phrase famously coined in 1948 by De Beers. However, in a large picture, there is no glamour and sophistication when it comes to the effects of environmental harm after mining a diamond and the weak welfare practices towards those who cultivate these stones. And as far as rarity is concerned, in fact 1% of the world's production of diamonds were produced in laboratories,[18] so that one speaks for itself.

4.2.2 Environmental Impact

Mined diamonds represent more than seven times the amount of effect relative to grown diamonds. The key factors that play a major role in the reason to distinguish how grown diamonds impact environmentally are.

Water and Energy Usage

The water usage is one of the largest area where mined and laboratory-grown diamonds vary. Mined diamonds absorb more than 126 gallons of water per carat. Lab-grown diamonds, on the opposite hand, absorb just 18 gallons.[19] Mined diamonds also contribute to continuous discharge of wastewater and contaminants into surface water bodies.

According to a research by Clean Origin,[20] energy-wise, 538.5 million joules per carat is often used by mined diamonds, while 250 million are being used by grown ones. Many of the energy that is used to produce laboratory-grown diamonds are renewable.

Whereas the energy sources used to run diamond mines produce greenhouse gases. In diamond mining, diesel fuels, electricity, and hydrocarbons all release harmful carbon into the air. Smog, climate change, and other environmental threats are caused by these substances.

16 TheVendeur [34].

17 Hailes [21].

18 Stone [32].

19 Osborn [29].

20 Clean Origin [13].

Carbon Emissions

The disparity between lab-grown and mined diamonds in carbon emissions is staggering. While for each single carat, a mined diamond emits more than 125 lb of carbon, lab-grown diamonds produce just 6 lb of carbon, a mere 4.8% of what mined diamonds generate, stated by the research of Clean Origin.[21]

Consequently, the same research also stated that more than 30 lb of Sulphur oxide is also generated by natural diamonds, while lab-grown diamonds emit none. In total, air emissions are 1.5 billion times higher on a single carat of mined diamond than on a laboratory-grown one. Therefore, the development process requires few to no pollution of significance in terms of overall gas emissions.

Land Disruption and Waste Generated

Almost 100 square feet of land is disrupted for every diamond carat mined using conventional techniques, and more than 5798 lb of mineral waste is made.[22] Mining further compensates for fragile ecosystem balances and leaves the soil unusable, long after the termination of mining operations.

But in contrast, only 0.07 square feet of land per carat and only 1 lb of mineral waste is disrupted by laboratory-grown diamonds.[23] According to Frost & Sullivan's study it is analyzed that diamond-growing facilities "*often are established in remote that have a minimal environmental impact and almost no impact on biodiversity in the area of operation.*"[24]

This risk has also been mentioned by the founder of world's first luxury jewelry brand to use lab-grown diamond: "Diamond mining has taken an enormous toll on environments around the world, nowhere more than Africa. We're talking huge, permanent holes dug into the earth, polluted ground waters, wildlife displacement—the list goes on."[25]–Laura Chavez, Founder of Lark & Berry

Human Impact Due to Incident and Disasters

Diamond mining is a hazardous occupation, and human rights abuses, forced labor and inhumane treatment of workers have also made the industry known. Mined diamonds often are also used as illicit currency to finance war and conflict-related operations all round in Africa. Diamond mines are often vulnerable to collapsing and explosion, and as a result of their occupations, employees also see an increased risk

[21] Clean Origin [13].

[22] Osborn [32].

[23] Osborn [32].

[24] Clean Origin [13].

[25] TheVendeur [34].

of cancer, hearing loss, respiratory problems, and other health problems. In 2010, for 69 days, 33 miners in Chile were trapped below ground in a collapsing mine.

By and large, extracted diamonds result in 1 accident per 1,000 employees per year while lab-grown diamonds result in none. The diamond mining sector also sees 80 days of missed work time (per 1,000 employees) due to injury every year, according to the study from Clean Origin.

Consequently, the study also revealed that on an average, mined diamonds are a result of 4.5[26](change to words) environmental impacts every year (events that either violate environmental laws or have a significant impact on local human, plant or animal life). Lab-grown diamonds have not yet resulted in any of these accidents and "Mining of diamonds poses a persistent threat to the surrounding environment," according to the study reported by Frost & Sullivan's. To put it simply, under controlled conditions, industries now have technology that poses no risk to human life and has a much smaller environmental impact.

4.2.3 Traceability

The lack of transparency is one of the primary concerns with natural diamonds. Even while purchasing potentially "ethical-sourced" diamonds for certain brands, there's just no way to tell whether that's the case.

The entire traceability concept is mainly driven by the demand from the consumers (millennial and gen-Z buyers) who are the present and future clientele. They are well informed by the trends in diamond industry and are keen in knowing history and provenance of the diamonds they are buying.

At present there are many existing traceability solutions, but they are insufficient. The most famous approach is Blockchain Technology that records declarations made by each party in the diamond supply chain. But even Blockchain cannot control a human error while recording data which leads to a long-term inexact data of legitimacy.

Then, there comes a procedure which collects this data separately from manufacturing systems during the manufacturing process, thereby reducing the opportunities for human falsification, and is then processed safely in the cloud. For example, this method is used by Sarine Diamond Technologies.[27] This is the best and only way available to guarantee absolute disclosure of the sources of diamonds that cannot be tampered with. There might be a never-ending process and unproven facts about how the diamonds are sourced but by instinct and depending on the situation, people will always remain responsible and distant when they think they are facing a counterfeit.

Here's what can be considered to look for in a brand when scouting for the genuine grade lab-grown diamond on the planet.

1. Look for brands that are committed to using renewable energies (some have lowered their carbon emissions to zero by doing so!).

[26] Willow [36].

[27] Sarine Diamond Technologies [31].

2. Check if they are buying offsets, if they are not using renewable energies.
3. Look for brands that recycle water or any other manufacturing materials or take measures to lower how much of these goods are being used.
4. Search for accountability and accuracy controls for all independent certifications.
5. Seek for facts beyond 'eco-friendly' and 'sustainable' arguments. Check the facts and statistics that the brands provide to let you know that they are truly ethical. Look for companies that hold "Certified Sustainable Diamond" certification provided by SCS Global Services.

The only difference in mined and grown diamond is their point of origins. Other than that, they are chemically, physically and optically same and indistinguishable. When this research speaks about the effects and impact of a lab-grown diamond in the ecology, it is not only about the ethical production but also about how the future of this industry can be preserved. Lab-Grown Diamond is already creating resonance in the gemstone industry.

4.3 Why Not Mined Diamonds?

Diamonds may be the girl's best friend, but will it ever be the Earth's best friend? Remember how geological pressure is triggered by natural diamonds. Well, diamonds must be under a large amount of rock for this pressure to occur knowing that the mines in which they are discovered are generally deep, even deeper than the height of Empire State building which is 443 m.

For miners, this means a huge amount of ground shifting. According to a study by Sustainable Jungle, each carat is linked up to 1,750 tons of earth movement. Using, some simple math and the fact that there are about 148 million carats of diamonds mined every year that means that 259 trillion tons of earth are being shifted every single year all for the sake of some jewel. These figures, and the fact that mine shafts can be observed from space like the Mir Mine in Siberia, make it simple to see all the deforestation and soil degradation caused by diamond mining.

Also, the study by Sustainable Jungle states that the carbon emissions per carat of extracted diamonds were two million times greater than those of lab-grown diamonds. Although this precise figure is open to some investigation, another report showed that CO2 emissions from diamond mining accounted to 1.5 million vehicles operating for the whole year.[28]

The Kimberley Process was structured to eliminate "conflict violence"—remember bombings, oppressive states, and massive quantities of political and sexual violence connected with the birthday or wedding present. However, the policy has mostly failed; there are flaws in the system that leaves it as a toothless protection for all those who are vulnerable. In simple terms, you never really know what you

---

[28] Sustainable Jungle [33].

get when it comes to natural diamonds. The velvet box may include some beautiful diamond studs—but it could also entail with war, slave labor, child labor, abysmal working practices, unequal wages, and human trafficking as well. As a matter of fact, these kinds of jewels are known as "Blood Diamonds" and they do not complement with anything.

To finish, distinctly lab-grown diamonds share same properties to every milligram as natural diamonds. Neither of the diamond is better than the other in terms of quality and authenticity. What matters most is the significance that the buyers put into the diamond and what it is meant to be.

The most beautiful flower in the world "Rose" is as real and carries the same emotion if it blooms wild every year in nature and if it is cultivated commercially. A diamond carries the same emotion and originality, whether it is wrested out from the earth by miners or cultivated in a lab by scientists.

5 Conclusion

Diamonds are exclusive and authentic. Their uniqueness ushers the history and romanticism with a feeling that is sparked by holding a piece of the ancient world. Lab-grown is a natural evolution of diamond with a technological metamorphosis. For every evolution, the concept of change takes time, however for Lab-grown diamond, it will not be the case. This research states that Lab-grown diamonds are going to be the new revolution in the diamond industry as there is a rapid consciousness on the environmental impacts and the concerns relating to the carbon footprint.

From the first attempts in the 50's, two different techniques have been identified as the major way to create diamonds in laboratory. Among them the HPHT (High pressure High temperature), which tends to generate the same condition of diamond making deep down in the earth crust, is largely implemented worldwide especially in China, United States and Russia with massive infrastructure to produce this contemporary resource. The second one is the CVD (Chemical Vapor Deposition) method which gives an alternative to HPHT growth process. Requiring less energy for the creation of the gemstone, the CVD process will give diamonds of smaller size but with striking qualities like the best high jewelry diamonds thanks to their intense purity. From a scientific point of view, diamonds are not even rare to find nowadays due to a large trading market provided by the overproduction of diamonds. Nonetheless, due to end in the sight of mining industry there will be inadequacy of mined diamonds in the near future.

In fact, lab-grown diamonds are limitless in theory. The arrival of grown diamonds has shaken the world of jewelry to its heart and caused diamond miners to fight back aggressively. Word of mouth, the spread of love stories and the increasing awareness of diamond appeal, were a powerful marketing tools that the luxury industry used. Diamond growers often mention that lab-grown are ethical and sustainable; whereas miners and their industry allies oppose that only gems plucked from the Earth should be contemplated as "precious" and "real." Some of these claims are arbitrary; others are only supported by sparse, self-reported, or industry-backed evidence. But it is not deterring anybody from manufacturing them.

Jewelers have different opinions about the stakes on laboratory grown. Some very traditional oriented companies such as Tiffany & Co are still reserved to lab-grown diamond. According to this article,[29] Andy Hart, Senior Vice President of Diamond and Jewelry Supply at Tiffany & Co said, "Our position is lab-grown diamonds are not a luxury material." On the other side, other innovative companies such as the French jeweler Courbet highlights laboratory diamonds as essential opportunity to disrupt modern jewelry and thus reduce the environmental impact of the mining sector.

This has now become a battle over image, and image is everything when it comes to diamonds. Meanwhile this research specifies that diamond mining would never be able to cope with lab-grown diamonds from an ethical point of view. And the mining industry should be careful trying to do so. It must be consciously kept in mind about the deceptive ecological image portrayed called greenwashing. Capitalism introduced it and it follows what the DPA is doing today, calling to a bid on whether to trust mined or lab grown diamond.

Also, this research specifies that there is a major inclination towards the harm in ecology, and the advancement in technology for sustainable environment has piqued our attention and curiosity. The major source is carbon footprint, which with its entire spotlight has differed to extremes in the process of acquiring mined diamonds and producing lab-ones. The DPA-commissioned study calculated that per cut, polished carat, and the average laboratory-grown diamond results in around 511 kg of carbon emissions. Using fuel and energy usage information supplied by mining firms, along with other greenhouse gas sources at the mining area, it was stated that the average mined carat was accountable with just 160 kg of carbon emissions.

One drawback here is that the calculation of the carbon footprint for mining reflects solely on the processing of diamonds, not the years of work involved in the construction of a mine. It can take a lot of energy to build a mine, particularly for those situated in remote areas where trucks or aircraft need to haul equipment for long distances. These also lead in impacting the environment and population around the mine area which is in comparison nowhere near to the minimizable carbon emissions released by the lab-grown diamonds.

Fortunately, several lab-grown diamond manufacturers have much to do with this. The primary goal of lab-grown diamond technology is to minimize the carbon footprint in the environment. Even if grown diamonds require huge amount of energy, some new improvements in technology emerged with less energy waste such as the use of HPHT system that stabilizes the polymers used for fluid loss control and rheology modification. And there are several industries that already have pledged or have already made improvements to increase their efficiency by using renewable energies, recycling methods, or carbon offsets.

Also, many manufacturers have already committed in providing the ethical supply-chain; the jeweler Courbet being one of them. This research showed that both source

[29] Blake [10].

of production of diamond, growing and mining, were consuming energy, and it is on great attention to focus on minimizing its consumption. These developments have shown a tremendous growth in the market, where the value for trust and credibility has been vouch by the facts and results shown by the lab-diamond producers. Also, according to the survey conducted by Instore in 2020, lab-grown diamonds are making their way to more than 52% independent jewelry houses who are stocking up the once controversial gems which is 10% more than four years ago.

Being the famous innovation in the gemstone industry and part of the sustainability trend, lab-grown diamonds turn out to face different risks, challenges, and opportunities according to the time period selected. The major concepts raised for the today's market are the disclosure and awareness of lab-grown diamond, which for many people stand as a question whether its real or fake.

The fierce competition with the established awareness of mined diamonds directly undervalues the effect that lab-grown diamond can usher. However, laboratory alternatives can be transposed as the market giants have already started to implement them. Indeed, the relevant value proposition included both Sustainability and Ethical reason leading the change in shift of the market towards this commodity. The future contingencies are oriented around the advancement in technologies and changing consumer behavior towards the ethically sourced diamond. Even though there are developments in the traceability process, there will always have challenges of accuracy.

Furthermore, if the diamond industry switches totally from mined to lab-grown which could be possible due to the shortage of raw gems in the mines, then the lab-grown industry should not only lead the new way of consuming but propose solutions to the developing economies that are dependent on mined diamond. For example, mined diamond revenues offer every child a free education in Botswana the second largest diamond producers in the world. Laboratory-grown industry should take deep actions to face the unemployment ratio and poor living conditions.

The opportunity to have the highest quality near-colorless diamonds in large quantities are interesting for many industries that depends on purity and excellence in accuracy like medical, high jewelry, electronics. The long-term vision of lab-grown diamond development will help to redefine the consumer buying habits, focusing more on quality rather than quantity depending on their needs and reducing the environmental impacts.

All in all, for every 30 years there is a change in the thought process; some might call it trend or some call it fashion, or some call it generation, but whatever may the term be the evolution which takes the generation forward is called a torchbearer and this generation is the Lab-Grown Diamond.

References

1. Boghossian S (2019) Dying luxury. Available via DIALOG. https://exjewel.com/2019/01/1-dying-luxury-please-help-me. Accessed 13 Nov 2020

2. Hood B (2019) Lab-grown diamonds are now a $280 Million business. In: Robb report. Available via DIALOG. https://robbreport.com/style/jewelry/lab-grown-diamonds-moment-2851840. Accessed 17 Nov 2020
3. Johnson R (2020) French jeweller using purely lab-grown diamonds. In: Politicallore. Available via DIALOG. https://www.politicallore.com/french-jeweller-using-purely-lab-grown-diamonds. Accessed 21 Nov 2020
4. Milman O (2020) Are laboratory-grown diamonds the more ethical choice to say, "I do"?. In: The guardian lifestyle. Available via DIALOG. https://www.theguardian.com/lifeandstyle/2020/mar/10/diamonds-lab-grown-climate-change. Accessed 17 Nov 2020
5. New Grown Diamond (2020) Why real synthetic diamonds are cost-effective?. New grown diamond register. Available via DIALOG. https://newgrowndiamond.com/blog/real-synthetic-diamonds. Accessed 12 Nov 2020
6. Saati G (2019) On the environmental & social impact of mined vs. synthetic diamonds. In: The diamond loupe. Available via DIALOG. https://www.thediamondloupe.com/sustainability/2019-03-11/environmental-social-impact-mined-vs-synthetic. Accessed 16 Oct 2020
7. Smetana A (2019) Lab-grown diamonds vs. mined diamonds: what you should know. In: Rogers&Hollands.com. Available via DIALOG. https://rogersandhollands.com/blog/post/lab-grown-diamonds-vs-mined-diamonds. Accessed 05 Nov 2020
8. A SG (2020) Lab grown diamonds are shaking up the industry. In: ba.sg. Available via DIALOG. https://ba.sg/lab-grown-diamonds-are-shaking-up-the-industry. Accessed 01 Nov 2020
9. Bates R (2019) An environmental expert on the ecological impact of lab-grown diamonds. In: Jckonline. Available via DIALOG. https://www.jckonline.com/editorial-article/eco-impact-lab-grown-diamonds. Accessed 17 Oct 2020
10. Blake E (2019) What will Tiffany & Co. Do about lab diamonds?. In: The engagement ring bible. Available via DIALOG. https://www.engagementringbible.com/what-will-tiffany-co-do-about-lab-diamonds/. Accessed 21 Oct 2020
11. Boghossian S (2020) Lab grown diamond report. In: ExJewel. Available via DIALOG. https://exjewel.com/insight-lab-grown-diamond-report. Accessed on 13 Oct 2020
12. Cai L; Qin C; Guo D;Yang H;Yi F; Zeng X (2020/13) Special-shaped powder pressure heating tube and synthesis block for diamond synthesis having excellent quality. CN Patent 111672420, 18 Sept 2020
13. Clean Origin (2020) Lab grown vs mined. Clean origin register. Available via DIALOG. https://www.cleanorigin.com/about-lab-created-diamonds/. Accessed 13 Nov 2020
14. Constable H (2020) The sparkling rise of the lab grown diamond. In: BBC Future Planet. Available via DIALOG. https://www.bbc.com/future/article/20200207-the-sparkling-rise-of-the-lab-grown-diamond. Accessed 14 Nov 2020
15. Danziger PN (2019) While mined diamond sales decline, The future of lab grown diamonds is much more than jewelry. In: Forbes. Available via DIALOG. https://www.forbes.com/sites/pamdanziger/2019/12/15/mined-diamond-sales-decline-as-the-future-is-in-laboratory-grown-diamonds/?sh=3b70968778d7. Accessed 14 Nov 2020
16. D'Arpizio C, Levato F, Prete F, De Montgolfier J (2020) Eight themes that are rewriting the future of luxury goods. In: Bain & Company. Available via DIALOG. https://www.bain.com/insights/eight-themes-that-are-rewriting-the-future-of-luxury-goods. Accessed 19 Oct 2020
17. Fair L, Frisby R (2019) The many facets of advertising diamonds with clarity. In: Federal Trade Commission. Available via DIALOG. https://www.ftc.gov/news-events/blogs/business-blog/2019/05/many-facets-advertising-diamonds-clarity. Accessed 06 Nov 2020
18. Frost L, Sullivan D (2014) Production of rough diamonds, environmental impact assessment: 16-31
19. Garside M (2020) Lab-grown diamond industry. In: Statista. Available via DIALOG. https://www.statista.com/topics/7108/lab-grown-diamond-industry. Accessed 14 Nov 2020
20. Graff M (2019) A quick review of the FTC's rules on lab-grown diamonds. In: national jeweler. Available via DIALOG. https://www.nationaljeweler.com/blog/7586-a-quick-review-of-the-ftc-s-rules-on-lab-grown-diamonds. Accessed 03 Nov 2020
21. Hailes S (2019) OPINION: Lab-grown diamonds are here to stay. In: Professional Jeweller. Available via DIALOG https://www.professionaljeweller.com/opinion-lab-grown-diamonds-are-here-to-stay Accessed 29 Nov 2020

22. Home Business Magazine (2020) How the Lab-Grown Diamond Market Is Shaking Up Business Model Changes Post-Pandemic. Home Business register. Available via DIALOG. https://homebusinessmag.com/businesses/business-spotlights/lab-grown-diamond-market-business-model-changes-post-pandemic. Accessed on 10 Nov 2020
23. Kolowich Cox L (2020) The 18 best advertisements & ad campaigns of all time. In: Hubspot. Available via DIALOG. https://blog.hubspot.com/marketing/best-advertisements. Accessed 12 Nov 2020
24. Li Y; Liao J; Tan D; Jun H; Zhang W; Wang J (2020/19) A kind of graphite tube heating body and diamond large single crystal synthetic chamber body [Machine Translation]. CN Patent 210875234, 30 Jun 2020
25. Linde O (2019) The Global Diamond Industry 2019- current perspective on the diamond industry plus a 50 year review. In: Bain & Company. Available via DIALOG. https://www.bain.com/contentassets/e225bceffd7a48b5b450837adbbfee88/bain_report_global_diamond_report_2019.pdf. Accessed on 10 Nov 2020
26. Little Switzerland (2019) Lab-grown diamonds. Little Switzerland register. Available via DIALOG. https://www.littleswitzerland.com/blogs/news/lab-grown-diamonds. Accessed 07 Nov 2020
27. Little Switzerland (2020) Lab-Grown vs. Mined Diamonds. Little Switzerland register. Available via DIALOG. https://www.littleswitzerland.com/blogs/news/lab-grown-vs-mined-diamonds. Accessed 03 Nov 2020
28. Matjuschkin V; Woodland A (2020/12) High-pressure synthesis of diamond. WO Patent 2020089204, 07 May 2019
29. Osborn J (2020) 3 Reasons You Should Buy Lab-Grown Diamond Engagement Rings. In MANOFMANY. Available via DIALOG. https://manofmany.com/lifestyle/reasons-you-should-buy-lab-grown-diamond-engagement-rings. Accessed 26 Oct 2020
30. Osman A (2018) Lab-grown diamonds are shaking up the world of fine jewelry. In: Vogue Business. Available via DIALOG. https://www.voguebusiness.com/technology/lab-grown-diamonds-debeers. Accessed 02 Nov 2020
31. Sarine Diamond Technologies (2020) The importance of traceability in the diamond industry. In: Sarine. Available via DIALOG. https://blog.sarine.com/importance-traceability-diamond-industry. Accessed 15 Nov 2020
32. Stone M (2019) Beyond the hype of lab-grown diamonds. In: Mining. GIZMODO. Available via DIALOG. https://earther.gizmodo.com/beyond-the-hype-of-lab-grown-diamonds-1834890351. Accessed 28 Oct 2020
33. Sustainable Jungle (2020) Lab-created diamonds Vs "Real" diamonds: such thing as ethical bling. Sustainable Jungle register. Available via DIALOG. https://www.sustainablejungle.com/sustainable-fashion/lab-created-diamonds. Accessed 13 Nov 2020
34. TheVendeur (2020) Lab grown diamonds vs mined diamonds—which is more ethical? Thevendeur register. Available via DIALOG. https://thevendeur.co.uk/lab-grown-diamonds-vs-mined-diamonds-which-is-more-ethical. Accessed 20 Oct 2020
35. Wilkinson TL (2018) Lab-Grown Diamond Jewlery Gains Sparkle. In: BILLIONAIRE magazine. Available via DIALOG. https://www.bllnr.com/innovation/lab-grown-diamond-jewellery-gains-sparkle. Accessed 16 Nov 2020
36. Willow F (2020) What are lab diamonds & Why they're more ethical and sustainable. In: Ethical Unicorn. Available via DIALOG. https://ethicalunicorn.com/2020/02/27/lab-grown-diamonds. Accessed 13 Nov 2020
37. Wu W (2019) Can Chinese millennials and gen zers warm the freezing global diamond industry? In: Jing Daily. Available via DIALOG. https://jingdaily.com/chinese-millennials-and-gen-zers-help-global-diamond-industry. Accessed 26 Oct 2020
38. Yukun L (2019) Lab-grown diamonds shaking up the industry. In: China Daily.com.cn. Available via DIALOG. https://www.chinadaily.com.cn/a/201910/05/WS5d97ff68a310cf3e3556ed43.html. Accessed 23 Oct 2020
39. Zhang B (2020/11) Artificial diamond synthesis device and synthesis method [Machine Translation]. CN Patent 111701535, 25 Jun 2020